Advances in Geographical and Environmental Sciences

Series Editor
Dr. R.B. Singh

For further volumes:
http://www.springer.com/series/13113

AIMS AND SCOPE

Advances in Geographical and Environmental Sciences synthesizes series diagnostigation and prognostication of earth environment, incorporating challenging interactive areas within ecological envelope of geosphere, biosphere, hydrosphere, atmosphere and cryosphere. It deals with land use land cover change (LUCC), urbanization, energy flux, land-ocean fluxes, climate, food security, ecohydrology, biodiversity, natural hazards and disasters, human health and their mutual interaction and feedback mechanism in order to contribute towards sustainable future. The geosciences methods range from traditional field techniques and conventional data collection, use of remote sensing and geographical information system, computer aided technique to advance geostatistical and dynamic modeling.

The series integrate past, present and future of geospheric attributes incorporating biophysical and human dimensions in spatio-temporal perspectives. The geosciences, encompassing land-ocean-atmosphere interaction is considered as a vital component in the context of environmental issues, especially in observation and prediction of air and water pollution, global warming and urban heat islands. It is important to communicate the advances in geosciences to increase resilience of society through capacity building for mitigating the impact of natural hazards and disasters. Sustainability of human society depends strongly on the earth environment, and thus the development of geosciences is critical for a better understanding of our living environment, and its sustainable development.

Geoscience also has the responsibility to not confine itself to addressing current problems but it is also developing a framework to address future issues. In order to build a 'Future Earth Model' for understanding and predicting the functioning of the whole climatic system, collaboration of experts in the traditional earth disciplines as well as in ecology, information technology, instrumentation and complex system is essential, through initiatives from human geoscientists. Thus human geosceince is emerging as key policy science for contributing towards sustainability/survivality science together with future earth initiative.

Advances in Geographical and Environmental Sciences series publishes books that contain novel approaches in tackling issues of human geoscience in its broadest sense – books in the series should focus on true progress in a particular area or region. The series includes monographs and edited volumes without any limitations in the page numbers.

Mehtab Singh • R.B. Singh • M.I. Hassan
Editors

Climate Change and Biodiversity

Proceedings of IGU Rohtak Conference, Vol. 1

Springer

Editors
Mehtab Singh
Maharshi Dayanand University
Rohtak, Haryana, India

R.B. Singh
University of Delhi
Delhi, India

M.I. Hassan
Ravenshaw University
Cuttack, Odisha, India

ISSN 2198-3542 ISSN 2198-3550 (electronic)
ISBN 978-4-431-54837-9 ISBN 978-4-431-54838-6 (eBook)
DOI 10.1007/978-4-431-54838-6
Springer Tokyo Heidelberg New York Dordrecht London

Library of Congress Control Number: 2014935232

Printed on acid-free paper

Springer is part of Springer Science+Business Media (www.springer.com)

Preface

The issue of climate change and biodiversity has become part of both the popular lexicon and the public discourse. Discussions on these subjects often evoke fierce debate between adherents to different views of the anticipated threat posed. Yet there are many nuances regarding climate change and the threats to biodiversity they represent that are not well understood by the public. Our conceptual understanding hinges largely on images and paradigms within the popular culture that are often little more than caricatures of the actual underlying scientific concepts. To appreciate the potential threat that climate change represents to the global society, it is necessary that we first understand the true science underlying this phenomena.

The Intergovernmental Panel on Climate Change (IPCC) in its recent report pointed out that the global average of land and ocean surface temperature data shows an increase of 0.85 (0.65–1.060 °C) over the period 1880–2012. The total increase between the average of the 1850–1900 period and the 2003–2012 period is 0.78 (0.72–0.85) °C. The atmospheric concentrations of the greenhouse gases, i.e., carbon dioxide (CO_2), methane (CH_4), and nitrous oxide (N_2O), have increased since 1750 due to anthropogenic activity. The overwhelming majority of climate scientists agree that human activities, especially the burning of fossil fuels, are responsible for most of the climate change currently being observed. These are very likely to trigger substantial changes in the structure and functioning of all ecosystems including biodiversity regions.

Climate change is already having an impact on biodiversity and is projected to become a progressively more significant threat in the coming decades. Climate change has a number of impacts on biodiversity, from the ecosystem to species level. Perhaps the most obvious effects are those that the changes in rainfall distribution, temperature, flooding, and sea level rise will have on ecosystem boundaries and the functions within them. Loss of Arctic sea ice threatens biodiversity across an entire biome and beyond. The related pressure of ocean acidification, resulting from higher concentrations of carbon dioxide in the atmosphere is also being observed. It is forcing biodiversity to adapt through changing habitat, life cycles, or development of new physical traits. This, in turn, will affect vital ecosystem services for all humans, such as air and water purification, pollination and production of food, decomposition and nutrient cycling, and carbon sequestration.

Biodiversity conservation can help to reduce the effects of climate change. For example, conservation of habitats can reduce the amount of carbon dioxide released into the atmosphere. If we act now to mitigate greenhouse gas emissions and identify systems-based adaptation priorities, we can reduce the risk of species extinctions and limit damage to ecosystems. We can preserve intact habitats, especially those sensitive to climate change; improve our understanding of the climate change-biodiversity relationship; and view biodiversity as a solution to climate change.

This book, consisting of 20 research papers presented at the IGU Conference, Rohtak, March 14–16, 2013, encompasses the interlinked issues of climate change and biodiversity in the above-mentioned interactive areas. Thus, the book aims to present a study of both natural and human realms and their climatic interactions, focusing on space and regions, addressing and questioning both short-term and long-term strategies. This work will be useful for students, researchers, and teachers in various disciplines such as geoinformatics, geography, climatology, meteorology, forestry, environmental studies, ecology, and biodiversity.

Rohtak, Haryana, India — Mehtab Singh
Delhi, India — R.B. Singh
Cuttack, Odisha, India — M.I. Hassan

Contents

Part I
Climate Change

Chapter 1
Climate Change and Glacial Lake Outburst Floods in Himachal Himalaya, India

R.B. Singh and Pankaj Kumar

Abstract The paper deals with relationship between climate change and glacial lake outburst in Himachal Himalaya, India. To qualify this objective primarily temperature and rainfall data of Bhuntar, Manali, Dharamsala and Mandi for a period of 30 years (1977–2007) has been obtained. A land surface temperature map has been derived using Landsat TM thermal band-6 to show changes in surface temperature between year 1989 and year 2011. Mean minimum temperatures of all stations show increasing trend with varying degrees ranging from 0.1 °C at Dharamsala to 2 °C at Mandi. In the Spiti valley, maximum surface temperature increase has been noticed between the heights of 4,000–5,000 m. The total annual rainfall for the state is 149 cm and the total annual number of rainy days is 65. Glacial lakes are a common feature at altitudes of 4,500–5,500 m in many river basins of the Himalaya. In order to identify potential Glacial Lake Outburst Floods (GLOFs), the present paper uses a technique adopted by Clague and Mathews (1973) for estimating maximum instantaneous discharge from a lake at the time of outburst. A total of 65 glacial lakes have been mapped in Chamba, Lahaul & Spiti and Kinnaur district, of which 23 have been identified as potential GLOFs.

Keywords Climate change • Glacial lake • Glacial lake outburst flood • Himalaya • Land surface temperature • Landsat TM

R.B. Singh (✉)
Department of Geography, Delhi School of Economics, University of Delhi, Delhi 110007, India
e-mail: rbsgeo@hotmail.com

P. Kumar
Department of Geography, Shaheed Bhagat Singh College, University of Delhi, Delhi 110017, India
e-mail: pankajdsedu@gmail.com

M. Singh et al. (eds.), *Climate Change and Biodiversity: Proceedings of IGU Rohtak Conference, Vol. 1*, Advances in Geographical and Environmental Sciences,
DOI 10.1007/978-4-431-54838-6_1,

1.1 Introduction

The global climatic change during the first half of the twentieth century has brought a tremendous impact on the high mountainous glacial environment. Villages at the foot hill of the Himalayan Mountain from Nepal to Bangladesh and to the whole length of India share a common history of natural disasters, many of which are spawned by the snow-white peaks looming above (Singh 2006). Glaciers in Himalaya are receding faster than in any other part of the world. In the last 100 years alone, the global mean temperature has increased by about 0.5–1 °C and the rapid receding of glaciers, to a major extent, is a consequence of global warming (Bahadur 2004). Many of the glaciers melted rapidly and gave birth to the origin of a large number of glacial lakes. Due to the faster rate of snow and ice melting the accumulation of water in these glacial lakes has been increasing rapidly and resulting into sudden discharge of large volume of water and debris and causing flooding in the downstream. Sudden, large river flow caused by an outburst of a glacier lake is generally termed Glacier Lake Outburst Flood (GLOF). GLOF causes disasters to life and property in the downstream region. In South Asia, particularly in Himalayan region, it has been observed that the frequency of occurrence of GLOF events has increased in the second half of the twentieth century. There is a need to monitor high altitude glaciated regions to understand the natural processes and the magnitude of natural hazards in order to promote effective mitigation measures. Accurate and timely information on the spatial location and regular monitoring of physical behaviour of glacier lakes is needed to monitor the GLOF hazards and assess the anticipated damages. Emerging geographical technology such as Remote Sensing and GIS can play a significant role in identifying potential GLOF risk together with monitoring it in real and near real time. This paper aims to study relationship between climate change and Glacial Lake Outburst and has identified and mapped potential GLOF sites in the Himachal Himalaya.

1.2 Study Area

Himachal Himalaya, situated in the lap of Western Himalaya between 30°22′44″ and 33°12′40″N latitude, and 75°45′55″–79°04′20″E longitude, and occupies an area of 5.57 million ha. It is bordered by Jammu and Kashmir on north, Punjab on the south-west, Haryana on south, Uttarakhand on the south east and China on the east (Fig. 1.1). It is a hilly state with a general increase in elevation from west to east and south to north ranging from 350 m to 7,000 m. Its one-third area remains snow covered for about 7 months in a year. This snowy part of the state is the source of three major rivers—the Beas, the Ravi, and the Chenab while the Satluj and the Yamuna Rivers originate from Tibet and Yamunotri, respectively. There are many

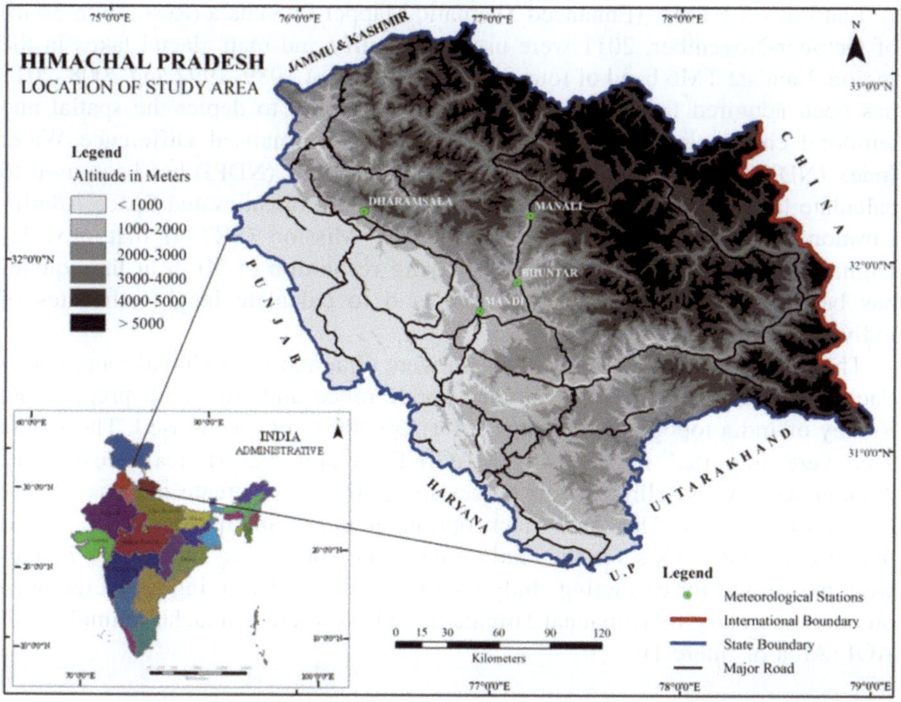

Fig. 1.1 Location and Altitudinal division of study area

glaciers and glacier lakes over the Zaskar and the Great Himalayan ranges. Himachal is also well known for its rich flora. Forests cover about 38 % of the state's total area. The general physiographic divisions from south to north are—The Outer Himalaya (Shivaliks), The Lesser Himalaya (Central zone) and The Great Himalaya (Northern zone).

1.3 Material and Methods

In this study, medium-resolution monthly average temperature and rainfall data prepared by the India Meteorological Department for the Himachal region from year 1973 to 2007 has been obtained for four meteorological stations of Bhuntar, Manali, Dharamsala and Mandi due to their close proximity to northern glacial environment. Annual Mean Maximum, Annual Mean Minimum and Annual Average Rainfall trend has been analyzed to show variability in climatic factors.

Landsat −7 ETM + (Enhanced Thematic Mapper Plus) data (Resolution-30 m) of October-November, 2011 were used to identify and map glacial lakes in the region. Landsat TM6 band of four consecutive year of 1989–1992 and 2008–2011 has been acquired for the month September–October to depict the spatial and temporal change in Land Surface Temperature. Normalised Difference Water Index (NDWI) and Normalised Difference Pond Index (NDPI) has been used to calculate lake area and other attributes. National Aeronautics and Space Administration (NASA) Shuttle Radar Topographic Mission (SRTM) that provides Digital Elevation Model (DEM) data having resolution of 90 m at the equator has been used to prepare altitude map and to calculate height attributes of individual lakes.

The satellite data covering study area were obtained from Global Land Cover Facility (GLCF) and Earth Explorer. As reference and base map preparation, Survey of India topographical sheets of 1:250,000 scales were used. These data sets were imported in ERDAS IMAGINE version 9.3 (Leica Geosystems, Atlanta, U.S.A.) satellite image processing software to create a False Colour Composite (FCCs). The layer stack option in image interpreter tool box was used to generate FCCs for the study areas. The sub-setting of satellite images were performed for extracting study area from images by taking geo-referenced out line boundary of Himachal Himalaya and Northern Himachal Himalaya as AOI (Area of Interest).

1.4 Result and Discussion

1.4.1 Climate Change and Mountain Ecosystem

Climate is an integral part of mountain ecosystems and organisms have adapted to their local climate over time. Climate change has the potential to alter these ecosystems and the services they provide to each other and to human society at large. There is a growing consensus in the scientific community that climate change is happening. While the absolute magnitude of predicted changes are uncertain, there is high degree of confidence in the direction of changes and the recognition that climate change effects will persist for many centuries. According to the Fourth Assessment Report of the Intergovernmental Panel on Climate Change, the global mean surface temperatures rose by 0.74 ± 0.18 °C over the last 100 years (1906–2005). The rate of warming over the last 50 years (0.13 ± 0.03 °C/10 year) is nearly twice that over the last 100 years (0.07 ± 0.02 °C/10 year) (IPCC 2007). In the Indian sub-continent, average temperatures are predicted to rise between 3.5 °C and 5.5 °C by 2100 (Lal 2002). Due to the increase in the average temperature in the mountain region, the rate of retreat of glaciers and the thawing of permafrost has rapidly increased in recent times. Thus, the region has witnessed unprecedented melting

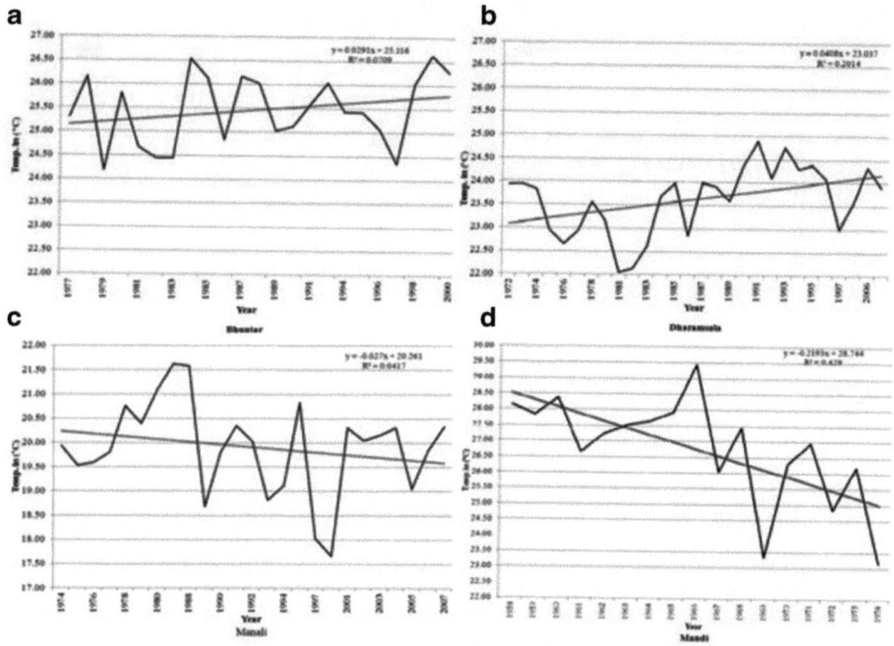

Fig. 1.2 Mean maximum temperature trend of (**a**) Bhuntar (**b**) Dharamsala (**c**) Manali and (**d**) Mandi stations of Himachal Pradesh (1977–2007)

of glaciers during the past three decades, with the vast Himalayan glaciers showing the fast rate of retreat, resulting in an increase in glacial runoff and glacial lake outburst floods (GLOFs).

1.4.2 Climate Trend Over Himachal Himalaya

1.4.2.1 Annual Trend in Mean Maximum Temperature

Annual temperature trend of Bhuntar meteorological station shows substantial increase of around 2 °C during 1977–2000 (Fig. 1.2). In the year 1977 the mean maximum temperature was 25.31 °C while it was 26.65 °C in 2007. Dharamasala meteorological station is also showing the same trend matching with Bhuntar station. About 2 °C increase in the mean maximum temperature has been calculated. The trend line plotted shows that in the year 1972 the initial point of trend line was around 23.1 °C while in the year 2007 it is at around 24.2 °C. Mean maximum annual temperature for the station Manali and Mandi is showing decrease during 1974–2007. Anthropogenic activities at local level have influence on such microclimate variability.

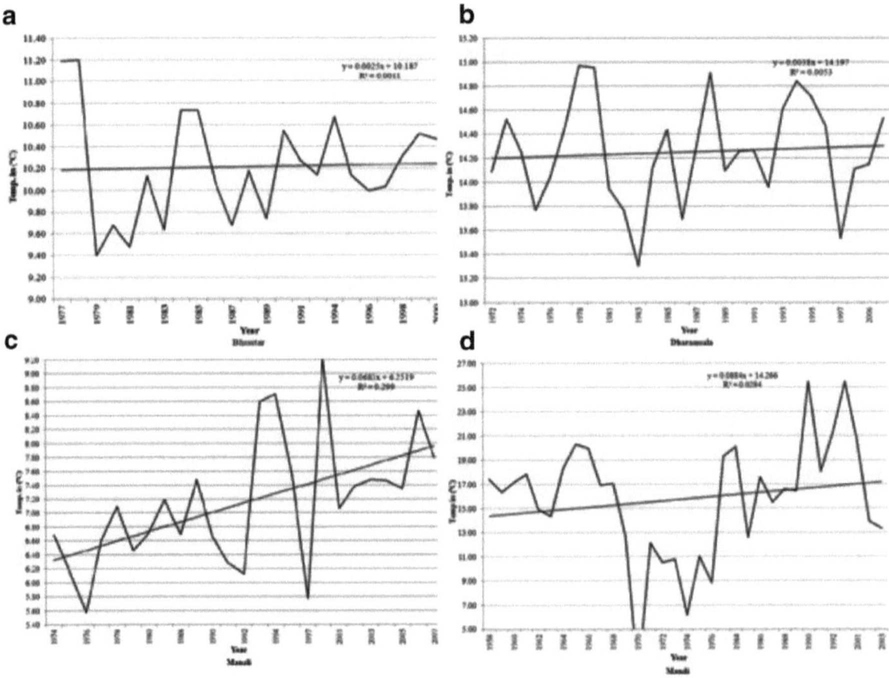

Fig. 1.3 Mean minimum temperature trend of (**a**) Bhuntar (**b**) Dharamsala (**c**) Manali and (**d**) Mandi stations of Himachal Pradesh (1977–2007)

1.4.2.2 Annual Mean Minimum Temperature

Mean minimum temperatures of all stations are showing increasing trend with varying degree ranging from 0.1 °C at Dharamsala to 2 °C at Mandi (Fig. 1.3). At Bhuntar meteorological station trend line projects an increase of around 0.3 °C in a time span of 30 years. Manali and Mandi depict comparatively higher increase of annual mean minimum temperature.

1.4.2.3 Land Surface Temperature Trend

The study is based on derived average values of imageries Landsat TM6 taken for four consecutive years of 1989–1992 and 2008–2011 for the month September–October to depict the spatial and temporal change. The range of the magnitude of land surface temperature in the region is ranging from −1.5 to 3.5 °C. Most of the area is showing an increase in the land surface temperature. The increase is visibly more pronounced in the Spiti valley, Kinnaur and in the northeast Chamba. In the Spiti valley, maximum surface temperature increase has been noticed between the heights of 4,000–5,000 m.

1.4.2.4 Annual Trend in Rainfall

The total annual rainfall in the region is maximum over the region of Kangra district and neighbourhood. The total annual rainfall for the state is 149 cm. Kangra district receives maximum amount of rainfall (185 cm), whereas Una receives the minimum amount of rainfall (121 cm) in a year. The rainfall over the state increases towards northeast region during winter and pre-monsoon. Annual mean rainfall trend plotted for Manali, Dharamsala, Bhuntar and Mandi does not show any clear picture of either decreasing or increasing trend. This proves notion of climatic uncertainty in the Himalayan region. Therefore, it is very difficult to derive clear cut linkages with altitude.

1.4.3 Climate Change and GLOF

Climate-change impacts are already occurring in the Greater Himalaya (Beniston 2003; Cruz et al. 2007). In recent times, the issue of concern for glaciologists and climate scientists has been the rate of retreat which has accelerated in the past few decades (Dyurgerov and Meier 2005). Glaciers and snow-pack are highly vulnerable to warming and shows the cyclical relationship. An increase in temperature causes snow to melt, which reduces the surface area of snow/ice, which further reduces the albedo of snow pack, which leads to absorption of heat energy, which reinforces melting (Knight 1999) resulting into the increased flow of water from snout of glaciers and formation of GLOF at terminal position. The shrinking of glaciers is accompanied by the formation of unstable glacial lakes that threaten downstream areas with outburst floods. GLOF is one of the most immediate and visibly dramatic effects of climate change in the Himalayan region (UNEP 2007).

1.5 Glacial Lake Outburst Flood in Himalaya

Glacial lakes are found behind moraine dams, on the surface of glaciers, within glaciers, and where glacial ice and/or a lateral moraine blocks a side valley (Ives 1986; Walder and Fountain 1997). Most of the glacial lakes in the Himalayan region are known to have formed within the last 5 decades, and a number of GLOF events have been reported in this region (Mool et al. 2001). On an average, in every 3–10 years one GLOF event was recorded in Himalayan region (Randhawa et al. 2005). These GLOF events have resulted in loss of many lives, as well as the destruction of houses, bridges, fields, forests and roads. Paree Chu lake outburst in the Spiti river in June 2005 has been covered widely by print and electronic media.

1.6 Identification of GLOF

Identification of Potential GLOFs in the present study uses technique adopted by Randhawa et al. (2005) considering the criteria of maximum instantaneous discharge from a lake at the time of outburst for the Satluj river basin. A simple empirical formula suggested by Clague and Mathews (1973) to estimate peak run-off was used in the investigation. In this method, the maximum instantaneous discharge can be estimated by using the volume as an input. The equation used for the estimation of peak runoff is:

$$Q_{max} = 75\left(\frac{V}{1000000}\right)^{0.67} \qquad (1.1)$$

where Q_{max} is the maximum instantaneous discharge (m^3/s); V is the lake volume (hm^3).

Estimation of lake volume has been done by using data of lake area and average depth of lake extracted from the Landsat TM of Sept–Nov, 2011. The lake area has been digitized from Landsat TM (30 m spatial resolution). Normalised Difference Pond Index (NDPI) has been used for the purpose. Maximum possible lake volume is important, as water level rises to maximum moraine dam lake can cause outburst flood.

$$V = \text{Average lake depth} \times \text{Lake area} \qquad (1.2)$$

The Geo Eye and SPOT image along with Digital Elevation Model (DEM) having high spatial resolution available on Google Earth has been directly taken for generation of cross-section profile of the lake. Randomly five glacial lakes have been selected to generate cross section profile and the average value derived has been taken as average depth for all other lakes in the region. Beside remote sensing data, Survey of India toposheet (1:250,000) and SRTM DEM having 90 m spatial resolution has been used for reference purpose. The calculation and analysis of data has been performed using Microsoft Excel and Arc GIS 10 along with ERDAS Imagine 9.3.

1.7 Potential GLOF in Himachal Himalaya

The number and the area of glacial lakes are increasing in the Northern Himachal in the past few decades posing threat to the people living downstream.

Therefore, it is imperative to find out the lakes posing potential threat in the near future. For identification of potentially dangerous glacial lakes, the glacial lakes associated with glacier dammed by lateral moraine or end moraine with an area larger than 20,000 m^2 have been considered and have been defined

Fig. 1.4 Glacier Lakes (**a**) Chandratal lake, near Batal , Lahaul valley, (**b**) Suraj tal lake, near Baralacha, Lahaul valley, (**c**) Nako lake, Spiti valley and (**d**) a new glacier lake near Darcha, Lahaul valley

as major glacial lakes by International Centre for Integrated Mountain Development (ICIMOD).

Northern Himachal comprises of three districts: Chamba, Lahaul and Spiti and Kinnaur. Most of the area of these districts lie above 3,500 m altitude and comprises the portion of Central Himalaya. Due to the suitability of climatic and physical conditions, this region possesses large number of glacial lakes (Fig. 1.4). A total of 17 glacial lakes have been identified in the Chamba district lying above 3,500 m altitude. Glacial lake no.14 which is an erosion lake is the largest among all comprising 43,030.97 m^2 of area while lake no. 3 is the smallest one. The peak discharge for glacial lake no.4, 14 and 17 has been estimated as 43, 67 and 56 m^3/s respectively. Mean Daily Discharge of river Satluj at Khab for the rainy seasons (June, July and August) for the period ranging from 1987 to 1999 come out to be 50 m^3/s. Taking the Mean Daily Discharge of 50 m^3/s as threshold for deciding a lake as potential threat, lake no. 14 and 17 of Chamba has been termed as potential GLOFs (Table 1.1). Out of the total 30 glacial lake in the Lahul and Spiti district, 22 qualified the criteria of 20,000 m^2 of areal coverage for further investigation of GLOF threat potential. Analysis of the result reveals the fact that out of total 22 qualified lakes in the district 16 have been identified as potential GLOFs (Fig. 1.5). Taking the above parameter glacial lake no. 4, 5, 7, 8 and 17 have been termed as potential GLOFs in the Kinnaur district (Fig. 1.5). These lakes are dangerous to the downstream population as their instantaneous water releasing capacity is more than the mean daily discharge of river Satluj. In Paree Chu lake

Table 1.1 Volume and Q_{max} of potential danger glacial lakes of Chamba, Lahaul & Spiti and Kinnaur district

Lake name	Longitude	Latitude	Area	Average depth	Volume (hm^3)	Q_{max} (m^3/s)
Cha_gl_14	76°40′18.71″E	32°55′48.64″N	43,030.97	20	860,619	67
Cha_gl_17	76°21′25.70″E	33°1′11.28″N	32,666.51	20	653,320	56
L&S_gl_8	77°11′42.70″E	32°45′43.73″N	65,510.32	20	1,310,200	90
L&S_gl_9	77°16′48.46″E	32°50′40.01″N	49,434.56	20	988,680	74
L&S_gl_10	77°19′45.97″E	32°43′22.93″N	40,182.65	20	803,640	65
L&S_gl_11	77°20′50.52″E	32°42′17.83″N	46,623.07	20	932,460	72
L&S_gl_13	77°30′7.16″E	32°47′30.94″N	54,668.90	20	1,093,360	80
L&S_gl_14	77°18′22.79″E	32°37′51.75″N	29,821.73	20	596,402	53
L&S_gl_15	77°13′06.25″E	32°31′36.57″N	778,535.38	20	15,570,700	472
L&S_gl_16	77°37′1.14″E	32°36′16.09″N	57,620.80	20	1,152,400	82
L&S_gl_17	77°32′49.18″E	32°29′54.13″N	1,169,571.64	20	23,391,432	620
L&S_gl_18	77°36′52.95″E	32°29′0.69″N	472,779.70	20	9,455,580	338
L&S_gl_20	77°26′50.96″E	32°14′43.23″N	47,176.11	20	943,520	72
L&S_gl_21	77°26′56.61″E	32°14′24.61″N	30,683.02	20	613,660	54
L&S_gl_25	78°16′17.13″E	32°21′48.18″N	91,228.59	20	1,824,560	112
L&S_gl_28	78°25′4.14″E	32°12′15.09″N	35,701.23	20	714,020	60
L&S_gl_29	78°29′15.77″E	32°8′56.68″N	46,414.34	20	928,280	71
L&S_gl_30	78°24′56.46″E	31°57′53.30″N	63,320.82	20	1,266,416	89
Kin_gl_4	78°41′10.71″E	32°1′18.02″N	42,146.00	20	842,920	66
Kin_gl_5	78°10′4.94″E	31°39′40.50″N	175,066.00	20	3,501,320	173
Kin_gl_7	78°15′9.81″E	31°20′22.35″N	94,018.00	20	1,880,360	115
Kin_gl_8	78°45′2.73″E	31°33′13.61″N	67,340.00	20	1,346,800	92
Kin_gl_17	78°41′51.63″E	31°13′31.63″N	55,095.00	20	1,101,900	80

Cha Chamba, *L&S* Lahaul Spiti, *Kin* Kinnaur, *gl* glacial lake
Source: Landsat TM

outburst in the Spiti river in June 2005, high silt content in the river resulted in the temporary shutdown of hydroelectric projects in the area. The washing away of bridges on the Spiti and the Satluj rivers affected 2,400 people living in 23 villages downstream.

1.8 Conclusion

The region possesses large number of glacial lakes due to the suitability of climatic and physical conditions. Variability of temperature and rainfall in the region further increases the magnitude of threat from these lakes. A total of 65 glacial lakes have been identified in the region, demands proper community initiative as well as policy measures to monitor potential threat. The present day risk for an outburst from glacial lakes occurring in Himachal Himalaya is moderate, but the risk of outburst of glacial lakes in the future could be anticipated high and it might occur in coming

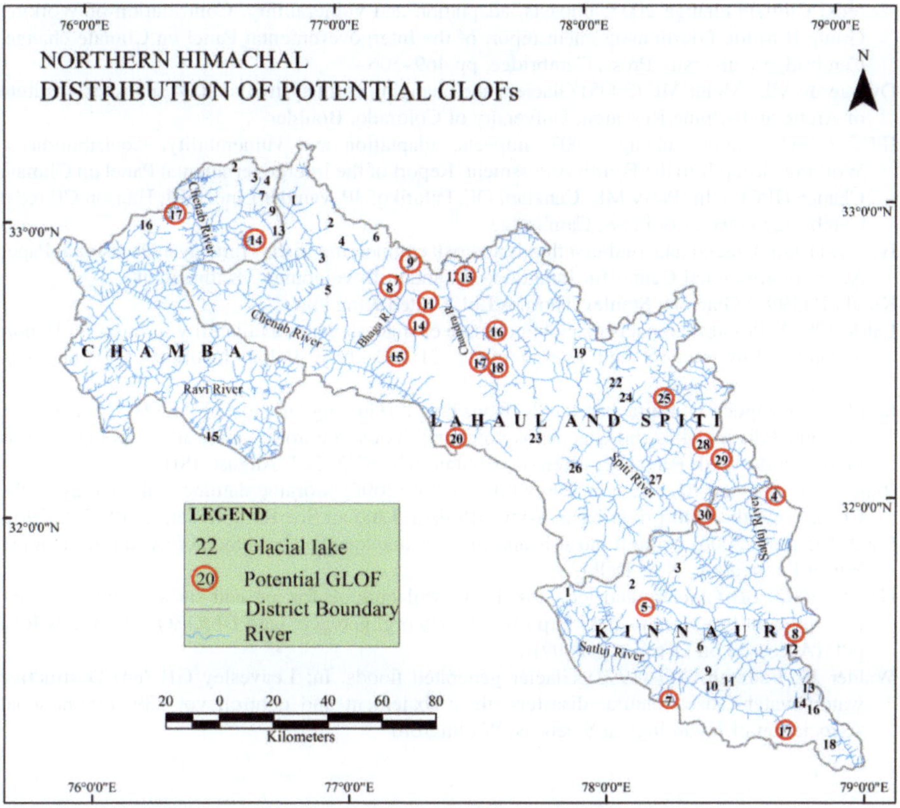

Fig. 1.5 Distribution of potential GLOFs in Himachal Himalaya

decades considering the present trend of climatic variability. Participatory GIS can be effective tool for inventorying glacial lakes, identifying its potentiality and mitigating the adverse effect in downstream region. Planners, policy makers, development workers, and scientists need to develop and implement appropriate mitigation measures including the implementation of early warning systems. Collaboration with local community is essential.

References

Bahadur J (2004) Himalayan snow and glaciers: associated environmental problems, progress and prospects. Concept Publishing Company, New Delhi

Beniston M (2003) Climatic change in mountain regions: a review of possible impacts. Clim Change 59:5–31

Clague JJ, Mathews WH (1973) The magnitude of Jokulhlaups. J Glaciol 12(66):501–504

Cruz RV, Harasawa H, Lal M, Wu S, Anokhin Y, Punsalmaa B, Honda Y, Jafari M, Li C, Huu Ninh N (2007) Asia. In: Parry ML, Canziani OF, Palutikof JP, van der Linden PJ, Hanson CE (eds) Climate change 2007: impacts, adaptation and vulnerability. Contribution of Working Group II to the Fourth assessment report of the Intergovernmental Panel on Climate change. Cambridge University Press, Cambridge, pp 469–506

Dyurgerov MD, Meier MF (2005) Glaciers and changing earth system: a 2004 snapshot. Institute of Arctic and Alpine Research, University of Colorado, Boulder

IPCC (2007) Climate change 2007: impacts, adaptation and vulnerability. Contribution of Working Group II to the Fourth Assessment. Report of the Intergovernmental Panel on Climate Change (IPCC). In: Parry ML, Canziani OF, Palutikof JP, van der Linden PJ, Hanson CE (eds) Cambridge University Press, Cambridge

Ives JD (1986) Glacial lake outburst floods and risk engineering in the Himalaya. Occasional Paper No. 5, International Center for Integrated Mountain Development, Kathmandu

Knight P (1999) Glaciers. Stanley Thornes Ltd, United Kingdom

Lal M (2002) Possible impacts of global climate change on water availability in India. In Report to Global Environment and Energy in the 21st Century. Indian Institute of Technology, New Delhi

Mool RK, Wangdev D, Bajrachaarya SR, Konzaang K, Guyong DR, Joshi SP (2001) Inventory of Glaciers, Glacial lakes and Glacial lake outburst floods-monitoring and early warning systems in the Hindu-Kush Himalayan Region, Bhutan, ICIMOD, 227, August, ISBN

Randhawa SS, Sood RK, Rathore BP, Kulkarni AV (2005) Moraine dammed lakes study in the Chenab and the Satluj river basin using IRS data. J Indian Soc Remote Sens 33(2):285–290

Singh RB (ed) (2006) Natural hazards and disaster management – vulnerability and mitigation. Rawat Publication, New Delhi

UNEP (2007) 6B Glaciers and ice caps. In: Global outlook for ice and snow. United Nations Environment Programme. p 38 http://www.unep.org/ geo/geo_ice/PDF/GEO_C6_B_LowRes.pdf. (Accessed on October 2, 2009)

Walder JS, Fountain AG (1997) Glacier generated floods. In: Leavesley GH (ed) Destructive water: water-caused natural disasters, their abatement and control, vol 239. International Association of Hydrological Sciences, Wallingford

Chapter 2
Climate Change Variability in Coastal Karnataka, India

R.B. Singh, Anju Singh, and Ashok Kumar

Abstract Climate change is one of the biggest environmental threats to food production, water availability, forest biodiversity and livelihoods. It is widely believed that developing countries such as India will be impacted more severely than developed countries. Global warming and Climate Change is projected to increase the number of extreme temperature and rainfall events, and hence climate variability is expected to show an upward trend. It is very important to understand the past trends and variability in rainfall, minimum and maximum temperature in Karnataka since the knowledge on the past could provide guidance for the future. The Arabian Sea and the North Kanara coast belong to the "Indo-Australian Marine Bio-geographic Region" considered to be the richest in the world for biodiversity. The coastal backwaters, estuaries, river-mouths are well known for their productivity. Some of them like the Aghanashini, Kali and Sharavati river backwaters, are even today so. The rivers from the Western Ghats carry great quantity of forest organic matter and deposit the same in the coastal waters including the sea. The current climate variability in Karnataka has been analyzed using the IMD daily rainfall data for the period of 1971–2005 and CRU data for the temperature for the period 1901–2002. The Costal Karnataka districts have an average rainfall of >25 mm/day. Decrease in precipitation trend has been observed in Coastal Karnataka. Rainfall has decreased by 17.69, 1.87 and 22.38 mm/day/100 year in Mangalore, North Kanara and Udupi respectively. The Increase in minimum

R.B. Singh (✉) • A. Kumar
Department of Geography, Delhi School of Economics, University of Delhi, Delhi 110007, India
e-mail: rbsgeo@hotmail.com

A. Singh
Department of Geography, Aditi College, University of Delhi, Delhi 110007, India

M. Singh et al. (eds.), *Climate Change and Biodiversity: Proceedings of IGU Rohtak Conference, Vol. 1*, Advances in Geographical and Environmental Sciences,
DOI 10.1007/978-4-431-54838-6_2, © Springer Japan 2014

temperature is 0.141, 0.146 and 0.111°C/100 year, whereas increase in maximum temperature is observed as 0.148, 0.146 and 0.113°C/100 year in districts of Mangalore, North Kanara and Udupi respectively.

Keywords Climate change • Climate variability • Coastal ecosystem • Environmental threats • India • Kanara coast

2.1 Introduction

The global climate is continuing to change at unprecedented rate in recent human history. The vulnerability of climate change and extreme climatic events such as drought and flood has dramatic impacts on economy and natural systems (Thornton 2006). Coastal regions are among the most vulnerable systems as it lies on boundary of terrestrial and marine ecosystems, which makes the climatic and environmental conditions in coastal areas very dynamic and sensitive too. The resilience and adaptive capacity of traditional networks and land use systems to cope with climate variability/extremes are weakening, while frequency and magnitude of climate variability and land use intensity are in rise.

Climatic variability is a complex concept including several issues of climate as temperature, rainfall, humidity and cloud conditions (Sen Roy and Singh 2002). A change in the variability of the climate is also considered as climate change even if average weather conditions could be as normal. India is known as one of the highly affected and vulnerable country to climate change owing to its large population depending on ecosystem services. Like other developing countries in India mostly population is actively engaged in activities which are more vulnerable, like agriculture, forestry, livestock and many more. People has lack of infrastructure, due to this even simple natural climatic variables turns in a severe climatic scenario. Further the means and capacity in developing countries adapt to changes in climate are scarce due to low level human and economic development and a considerable high proportion of poor population in the society (Downing et al. 1997).

The impact of climate change on the basic sectors decides the future food security of human being on the earth surface. The dimensions of the climate change and ecosystems are changing continuously. During the earlier times, ecosystem was considered to bear the bad impact of climate change (Sen Roy et al. 2011). The climatic sensitivity of coastal zone is not certain because of variability in rainfall, temperature and other climatic phenomena. The inter-annual variation in these factors of climate is increasing. Weather predictions of either short time or long time go out wrong and due to wrong predictions the economic loss used to increase (IPCC 2001; Sahu et al. 2012, 2013).

Coastal zones have certain aesthetic and practical advantages and this has resulted in the concentration of residential, recreational and industrial areas in this zone. Nearly two thirds of the World's population live in these areas which are often only one or two meters above the mean sea level. Most of these coastal areas are geologically young and are still being shaped by nature. People have chosen to live close to the water front either ignorant of or ignoring deliberately the vulnerability involved, the vulnerability that this boundary is not stable and not clearly defined, Water gives and takes and a state of dynamic equilibrium exists, More often the fluctuations that are part of this dynamic equilibrium are underestimated leading inevitably to disaster, loss of life and loss of property. This fragile dynamic equilibrium can often be upset by human interference, it becomes necessary that this coastal zone is managed and developed in a way that respects its natural and social significance, which ultimately lead to change in coastal climate and problems for livelihood.

The coastal zone of Kanara is one of the highly urbanised and better-developed geographical areas of the State with high degree of economic development and density of population. The settlements in the coastal region consist of 22 urban agglomerations. The occupational pressure by these urban areas can be attributed to fish landing and processing, port maintenance, mining for lime shell, bauxite and silica sand and coir retting. The occupational pressure is likely to be increased in the urban areas of Mangalore and Udupi regions, which will lead to further growth of urban population and thereby increasing the vulnerability posed by climate change. Therefore the present paper aims to assess the spatio-temporal extent and magnitude of climate variability in the coastal zone of Kanara.

2.2 Study Area

Kanara coast is located between 12°27′ to 15°32′N latitude and 74°05′ to 75°45′E longitude, its geographic area is 18,732 km^2, the coast stretches for 320 km along the three districts of South Kanara, Udupi and North Kanara. Of these North Kanara has 160 km long coastline while 98 km are in Udupi district and the rest in South Kanara. There are three distinct agro-climatic zones ranging from coastal flatlands in the west with undulating hills and valleys in the middle and high hill ranges in the east that separates from the peninsula. There is a narrow strip of coastal plains with varying width between the mountain and the Arabian Sea, the average width being about 20 km. The average height of the hinterland is 70–75 m, but in some places it can be as high as 150 m. Fourteen rivers drain their waters into the shore waters of Kanara coast (Singh 1997).

The three districts consist of 19 Talukas of which 8 are coastal talukas. In which 5 are in North Kanara, 2 are in Udupi and 1 is in South Kanara district. Kannada, Konkani and Tulu are the major Regional Language spoken. As per 2001 Census, the total population of coastal districts is 4,363,617,with average density of 278 km^2 (Census of India 2011) (Table 2.1). There are about 90 beaches with varying aesthetic potential that are suitable for beach tourism.

Table 2.1 Socio-Economic Profile of Kanara Coast

Geographic area	18,732 km^2	Adult literacy rate	62.41 %
Total population	4,359,196	Life expectancy at birth (both sexes together)	67 years
Proportion of female population	47 %	Life expectancy at birth of females	70 years
GDP per capita income	9,742	Total infant mortality	49/1,000
Population below poverty line	25 %	Mean age at marriage of females	20.89
Households without civic amenities (Drinking water, electricity and toilets)	33.5 %	Sex ratio	970/1,000

Source: Human Development in Karnataka 2005

2.3 Research Methodology

Both primary as well as secondary data sources are used in the present study. The temporal and spatial analysis has been done by using following techniques:

2.3.1 Moving Average (Five Years)

This tool is used to find out the trend of temperature and rainfall variation over the years.

2.3.2 Water Balance (FAO and Thornthwaite and Mather)

In this study whenever the monthly rainfall crosses 50 % Potential Evapotranspiration (PET), the length of growing period starts and the season gets terminated when the monthly rainfall is lesser than 50 % PET. The comparison between this water balance and previous one is helpful in determining the change.

2.3.3 Thornthwaite's Moisture Index and Climatic Classification

The classification of the climate is the appropriate way of analyzing climatic conditions of a region. Temperature and rainfall are the two main aspects of climatic classification. Thornthwaite's 1948 climatic classification was based on the evapotranspiration and vegetation index. Evaporation is the total amount of

Table 2.2 Thornthwaite's humidity provinces based on moisture index

Moisture index	Humidity provinces
100 and Above	Pre-humid (A)
80–100	Humid (B4)
60–80	Humid (B3)
40–60	Humid (B2)
20–40	Humid (B1)
0–20	Moist Sub-humid (C2)
−33.3–0	Dry Sub-humid (CI)
−66.7 to −33.3	Semiarid (D)
−100 to −66.7	Arid (E)

moisture which is evaporated from water bodies and transferred to atmosphere with amount of transpiration from living organisms. It is very difficult to call a climate, moist or dry, only by measuring level of precipitation; rather it is essential to know whether the precipitation is more or less than the water needed for evaporation and transpiration (Thornthwaite 1948). The potential evapotranspiration is calculated through mean monthly temperature (°C) with corrections for day length for a 30 day month.

$$\mathrm{PE(cm)} = 1.6(10\mathrm{t}/\mathrm{I})^{\mathrm{a}}$$

where I is the sum of 12 months of $(t/5)^{1.51\ 4}$; a is the Further complex function of I.

The monthly water surplus (S) or deficit (D) is calculated a moisture budget assessment including stored soil moisture. A moisture index is given by the following formula is as:

$$\mathrm{IM(100S\text{-}60\ D)\ PE\ or\ 100(P\text{-}PE)/PE}$$

The most important feature of this classification is that the temperature efficiency is calculated from PE value which is function of temperature. Using computed indices of moisture and heat, Thornthwaite defined the humidity provinces based on the moisture index (Table 2.2).

2.4 Results and Discussion

2.4.1 Temperature Variability

The wind direction over the Arabian Sea shows dramatic changes in temperature. The months of June and September, a time generally referred to as the summer monsoon, the general direction of winds over north of the equator is south-westerly and its strength is significantly larger than that during the rest of the year.

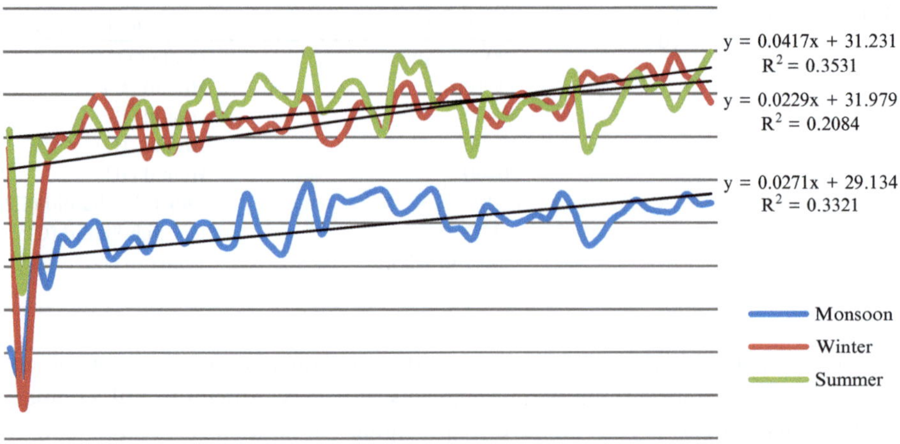

Fig. 2.1 Trend of Mean maximum temperature during 1953–2010 in Udupi district

During November to March, referred to as the winter monsoon, winds over the same region have overall north easterly direction. October and April–May are times of transition between the two monsoons causes variations in temperature too (Shetye et al. 1985).

2.4.1.1 Mean Maximum Temperature

The climate of coastal region is mainly influenced by land and sea breeze. The spatial variations in climate are also influenced by topography of the region. Out of the three study districts, North Kanara is having highest area under hills. The districts of Udupi and South Kanara are mostly coastal plain. The phenomenon of climate change is also observed in the region and variations are found in extent and magnitude of phenomenon. The change in mean maximum temperature is found to be highest in summer month (0.03°C per year) in Udupi district during the period of 1953–2010, with a coefficient of variation of 10 % (Fig. 2.1). The overall annual maximum change in mean maximum temperature is observed for the North Kanara station (Fig. 2.2). The minimum variations have been observed in South Kanara district for the month of monsoon (0.016°C per year) (Table 2.3).

2.4.1.2 Mean Minimum Temperature

The temperature dips to the lowest in morning around 3 o'clock. The mean minimum temperature has a considerable impact of cloud cover also. In the region the maximum change in minimum temperature is observed in the North Kanara district for the summer month (0.28°C per year) due to its

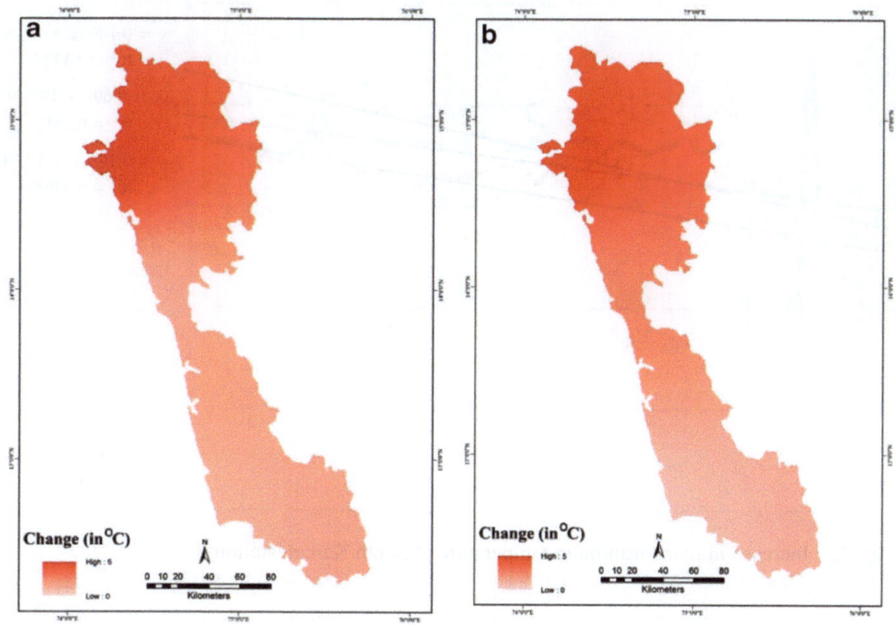

Fig. 2.2 Spatial pattern of change in mean maximum temperature during (**a**) summer season and (**b**) winter season

Table 2.3 Season wise variations in mean maximum temperature at three stations

	Monsoon (June–Oct)	Winter (Nov–Feb)	Summer (March–May)
North Kanara			
Average	29.480	31.945	32.172
Standard deviation	0.870	2.356	3.000
Coefficient of variation	2.951	7.375	9.327
Udupi			
Average	29.919	32.441	32.643
Standard deviation	0.779	1.165	0.833
Coefficient of variation	2.604	3.592	2.552
South Kanara			
Average	29.344	32.682	33.656
Standard deviation	0.844	0.745	0.583
Coefficient of Variation	2.878	2.281	1.734

undulating topography and proximity to Goa urban and industrial area (Fig. 2.3). The minimum change is noticed in Udupi station for monsoon month (0.05°C per year). The coefficient of variation is highest for the summer month in North Kanara station (Table 2.4).

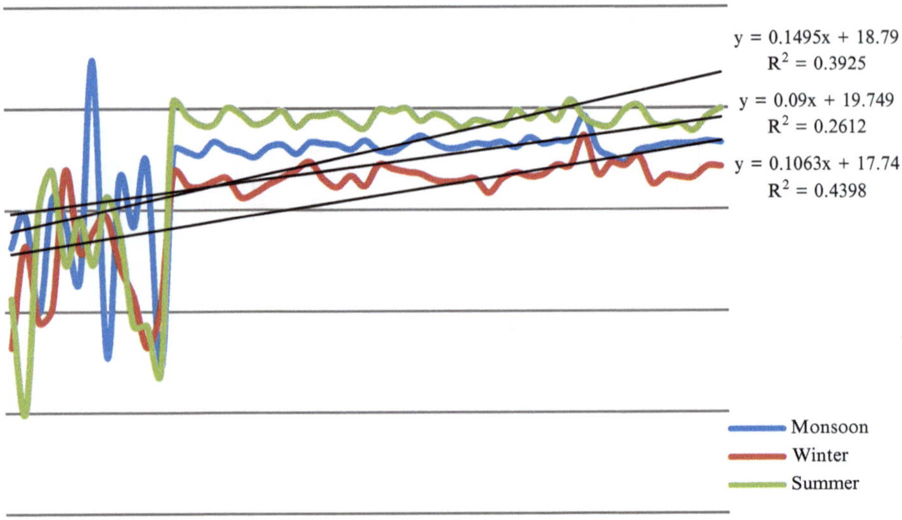

Fig. 2.3 Increase in mean minimum temperature at South Kanara station

Table 2.4 Season wise variations in mean minimum temperature at three stations

	Monsoon (June–Oct)	Winter (Nov–Feb)	Summer (March–May)
North Kanara			
Average	23.314	18.938	20.846
Standard deviation	5.425	3.457	6.694
Coefficient of variation	23.270	18.255	32.114
Udupi			
Average	23.134	20.000	21.350
Standard deviation	7.067	2.965	5.566
Coefficient of variation	30.547	14.826	26.073
South Kanara			
Average	22.177	20.610	22.827
Standard deviation	2.718	2.475	3.685
Coefficient of variation	12.258	12.011	16.146

2.4.2 *Rainfall Variability*

The study area lies west to the Western Ghats having per-humid climate according to Thornthwaite's climatic classification. About 90 % rainfall occurs in monsoon season. Maximum rainfall in monsoon season occurs at Udupi station (693 mm). In summer season maximum rainfall has been observed in South Kanara station (72 mm) (Fig. 2.4). The influence of climate change on rainfall has been maximum

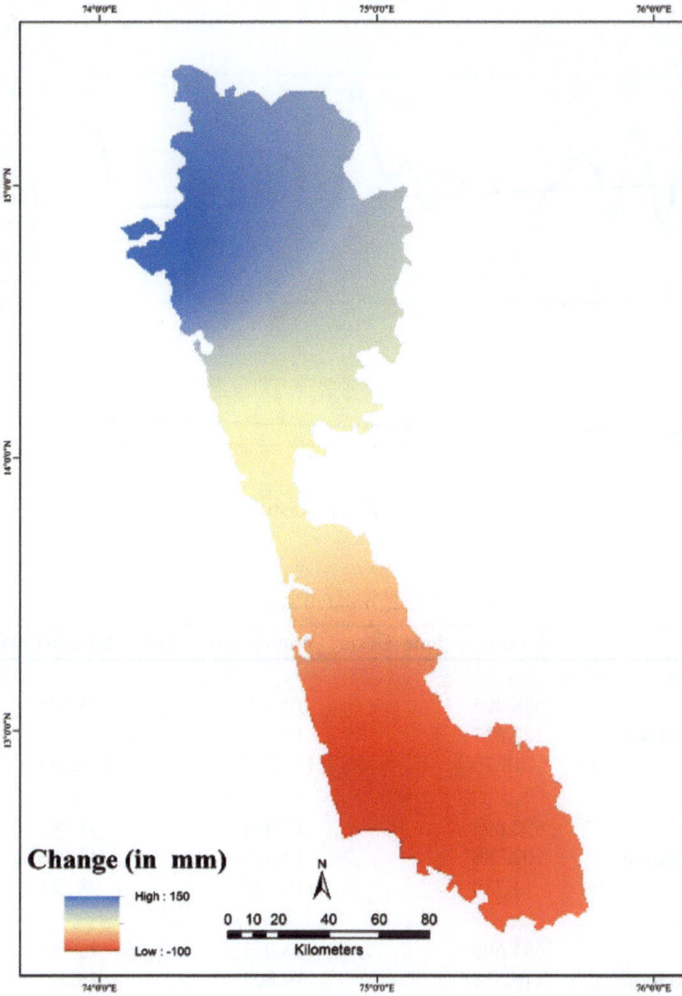

Fig. 2.4 Spatial pattern of change in average rainfall during monsoon season between 1953 and 2010

in South Kanara district. The rainfall in South Kanara district has reduced at the rate of 1.62 mm/year (Fig. 2.5). The other two stations have experienced an increase in rainfall during 1953–2010. The variation in rainfall is maximum for North Kanara district in winter season, the value of coefficient of variation is 149 %. The minimum variation is found in Udupi district for monsoon month with only 14 % coefficient of variation (Table 2.5).

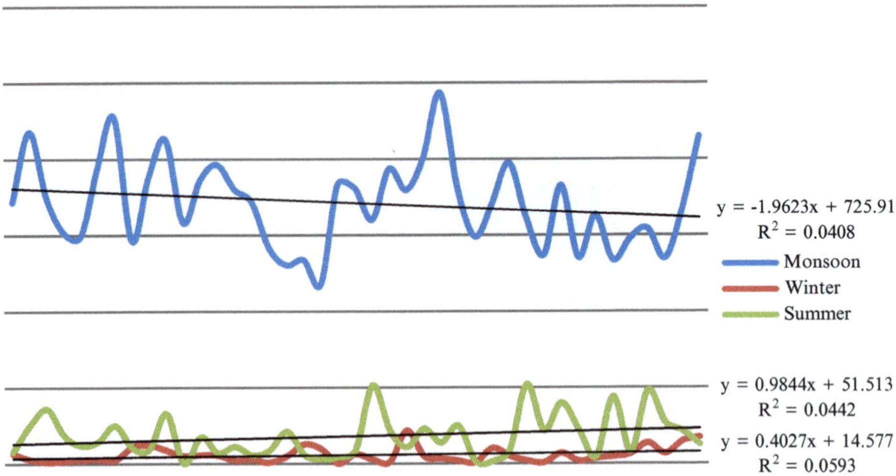

Fig. 2.5 Change in rainfall pattern in South Kanara district

Table 2.5 Season wise variations in average rainfall at three stations

	Monsoon (June–Oct)	Winter (Nov–Feb)	Summer (March–May)
North Kanara			
Average	588.495	12.460	45.804
Standard deviation	127.527	18.597	55.141
Coefficient of variation	21.670	149.248	120.385
Udupi			
Average	692.693	14.684	47.788
Standard deviation	102.386	17.381	46.986
Coefficient of variation	14.780	118.362	98.321
South Kanara			
Average	684.696	23.033	72.185
Standard deviation	116.318	19.802	56.074
Coefficient of variation	16.988	85.972	77.680

2.4.3 Change in Moisture Index

The trend of increasing temperature and changing precipitation scenario caused the shift in humidity provinces in coastal region of Karnataka. The analysis of correlation matrix indicates that the change in potential evapotranspiration which further causes the change in humidity provinces are caused by change in maximum temperature. Though there are significant changes in humidity indices throughout the coastal region of Karnataka but North Kanara district has observed maximum decrease in moisture index. By applying the Thornwaite's method of identifying and classifying regions into appropriate climatic category it is found

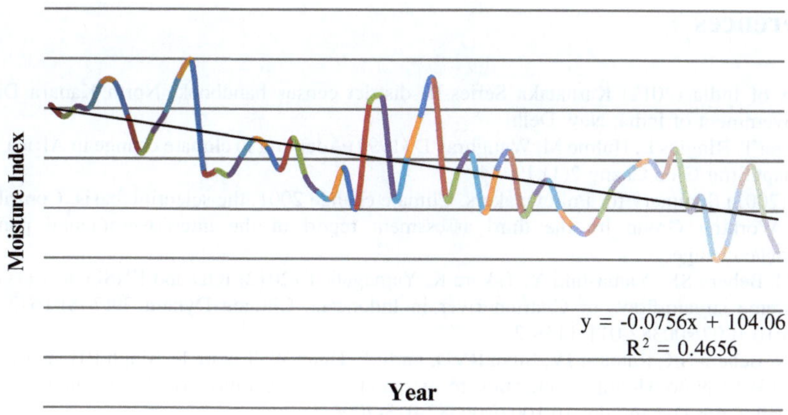

Fig. 2.6 Change in Thornthwaite's moisture index during 1953–2010

the coastal region of Karnataka has been transformed from per-humid to humid category due to combined effect of expansion of urban and industrial area. The moisture index has decreased from average value of 104 in 1953 to 98 in 2010 (Fig. 2.6).

2.5 Conclusion

Climatic variability is the most significant phenomenon affecting almost all economy sectors of the region. Variability also poses problems before fishermen and farmers to adopt certain measures as they may become useless after one cycle of variability. As spatial variability there is also temporal variability in climatic parameters. Rainfall in coastal region is concentrated in only 4 months of rainy season and variability leads to the fishermen and farmers to take wrong measures. To cope with the problem, it is essential to assess in advance the trend of future climatic variability based on past experiences.

The variability in the average temperature for all three seasons of summer, winter and monsoon is diverse in the region. In summer season the region is experiencing more temperature rise while in monsoon season less temperature rise is experienced. The precipitation has a decreasing trend in Udupi district whereas other two districts have experienced increase in precipitation. Number of wet days has also decreased over the years and the season of monsoon rainfall has been shrinking.

Acknowledgement Authors are grateful to Mr. Ajay Kumar for his assistance in preparation of the paper.

References

Census of India (2011) Karnataka Series-II: district census handbook, North Kanara District. Government of India, New Delhi

Downing T, Ringius L, Hulme M, Waughray D (1997) Adapting to climate change in Africa. Mitig Adapt Strat Glob Chang 2(1):19–44

IPCC (2001) Summary for Policymakers, Climate change 2001: the scientific basis. Contribution of Working Group Ito the third assessment report of the intergovernmental panel on climate change

Sahu N, Behera SK, Yamashiki Y, Takara K, Yamagata T (2012) IOD and ENSO impacts on the extreme stream-flows of Citarum river in Indonesia. Climate Dynam 39(7–8):1673–1680. doi:10.1007/s00382-011-1158-2

Sahu N, Behera SK, Ratnam JV, Silva RVD, Parhi P, Duan W, Takara K, Singh RB, Yamagata T (2013) El Niño Modoki connection to extremely-low streamflow of the Paranaíba River in Brazil. Clim Dynam. doi:10.1007/s00382-013-2006-3

Sen Roy S, Singh RB (2002) Climate variability, extreme events and agriculture productivity in mountain regions. Oxford and 1BH Publication, New Delhi

Sen Roy S, Singh RB, Manoj K (2011) An analysis of local spatial temperature patterns in the Delhi Metropolitan area. Phys Geogr 32:114–138

Shetye SR, Shenoi SSC, Antony AK, Kumar VK (1985) The Monthly-mean wind stress along the Coast of the North Indian Ocean. J Earth Syst Sci 94:129–137

Singh RL (1997) India: a regional geography. UBS Publications, New Delhi

Thornthwaite CW (1948) An approach towards a rational classification of climate. Geogr Rev 38(1):55, June

Thornton PK, Jones PG, Owiyo T, Kruska RL, Herrero M, Krisjtjanson P, Notenbaert A, Bekele N, Omollo A, Orindi V, Ochieng A, Otiende B, Bhadwal S, Anantram K, Nair S, Kumar V, Kelkar U (2006) Mapping climate vulnerability and poverty in Africa. Report to the Department of International Development, ILRI, Nairobi. Online at http://www.dfid.gov.uk/reserach/mapping-climate.pdf

Chapter 3
Assessment of Climate Induced Soil Salinity Conditions of Gosaba Island, West Bengal and Its Influence on Local Livelihood

Anwesha Haldar and Ajay Debnath

Abstract Climatic variability and occurrence of extreme climatic events are a major concern of the Indian Sundarbans. The impact of climate change could hinder development and delay the processes in eradicating poverty, hence potentially aggravating the social and environmental conditions in these areas. Crop productivity has either become static or shown a decline in some locations. Ecological degradation due to salinization has also affected the types and productivity of the crops and freshwater fishes. It is understood from the tonal variation in multi-dated satellite images of the area affected by Aila, a severe cyclonic storm that occurred on 25th May, 2009, that the ingress of saline water on agricultural land and subsequent infiltration and evaporation of water has left large areas fallow. Food crops have to be produced on the shrinking land resources as few suitable lands are available for cultivation. The perceived rising temperatures and uncertainties in rainfall may have serious direct and indirect consequences on crop production and hence food security of this region. At this point, innovations in the type of crops and process of cropping are of utmost necessity. This paper highlights the geographical impacts of Aila on the agriculture, occupation and life of the people in Gosaba Block in the Indian Sundarbans.

Keywords Climate change • Crop productivity • Indian Sundarbans • Soil salinization

A. Haldar (✉) • A. Debnath
Department of Geography, University of Calcutta, Kolkata, India
e-mail: anwesha.h.5@gmail.com; ajaydebnath89@gmail.com

M. Singh et al. (eds.), *Climate Change and Biodiversity: Proceedings of IGU Rohtak Conference, Vol. 1*, Advances in Geographical and Environmental Sciences,
DOI 10.1007/978-4-431-54838-6_3, © Springer Japan 2014

3.1 Introduction

Climate change is a long term shift or alteration in the climatic regime of a specific location, a region or the entire planet leading to environmental hazards and change in the human way of life. It is manifested in the coastal regions through sea level changes, salinization of fresh water sources, frequent occurrences of destructive storms and cyclones, increased variability of monsoonal rains and temperature conditions, fall in groundwater levels, increased pest population and resultant crop losses. Such extreme events have taken the greatest toll on the tropical countries like India due to their weak adaptive capacities, lack of timely financial support and high dependence on climate sensitive sectors like agriculture, marine and forest resources.

Agricultural productivity is sensitive to two broad classes of climate-induced factors. Firstly, the direct effects from changes in temperature, precipitation, radiation or carbon dioxide concentration; and secondly, the indirect effects through changes in soils and the distribution and frequency of infestation by pests and diseases (Mitra et al. 2010). The vulnerability of agricultural production to climate change depends not only on the physiological responses of the affected plant, but also on the ability of the affected socio-economic systems of production to cope with changes in yield, as well as with changes in the frequency of droughts or floods. The adaptability of farmers is severely restricted by the heavy reliance on natural factors and the lack of complementary inputs and institutional support system. Agricultural production in India has high vulnerability to current inter-annual climatic fluctuations (Kalra et al. 2003). Changes in soil and crop processes, growth, yield and cropping system have been detected in regions exposed to climatic vulnerabilities. Hence the study of crop responses to the biotic and a-biotic stresses is of utmost necessity for food security and maintaining the ecological balance.

3.2 Objectives of the Study

There has been much debate on actual crop productivity scenario in the Sundarban region during the last few years. The severe cyclone, *Aila* on May 25th, 2009 caused high storm surges, coastal inundation and heavy flooding, leaving the soil saline thereafter. The main objectives of this study is to establish the fact that the existing crop type and pattern are unsuitable in the present salinity and pH conditions of the soil and highlight the present socio-economic condition of the people. Thus the ground situation is being presented whereby it can attract innovations in crop seeds and cropping pattern for an adequate yield. This in turn can increase their income levels and assure them a better livelihood.

3.3 Methodology

The work has been done through the various steps of gathering information from books, articles, journals and social media to identify the problem areas. A pilot survey and subsequent field surveys in the villages of Gosaba Island have been carried out for water and soil sample collection and household questionnaire survey on their experiences and perceptions in agriculture in that area through a random sampling method. This was followed by testing and analysis of samples, data processing in MS-excel and mapping by Remote Sensing and GIS software-salinity and pH mapping in Surfur 10, Mapinfo 10 and Geomatica 9.1. Finally the information collected was tabulated for interpretation and reporting.

3.4 History and Geographical Setup of the Study Area

The formation of the Bengal delta is said to have initiated some 125 million years ago. After the 'Proto-Delta' building stage, (Sanyal 1999) the seaward regression of land ceased and sedimentation phase continued up to 4,000 year B.P. when the land took its present shape (Khan and Hoque 2002; Islam and Gnauck 2008).

The estuarine system of the Ganga- Brahmaputra deltaic region forms the largest mangrove eco-system in the world shared between India and Bangladesh approximately in the proportion of 60:40. Sir William Wilson Hunter in his 'A Statistical Account of Bengal' (1875) writes, "*The Sunderbans may be described as a tangled region of estuaries, rivers and water courses, enclosing a vast number of islands of various shapes and sizes*" (Hunter 1875). The extension of the region is roughly between 21° 30′ N to 22° 30′N and 88° 10′E to 89° 10′E covering about 104 islands of which 54 have been completely deforested. The straight line length of the coastline in the Indian Sundarbans from Sagar to Herobhanga at the Indo-Bangladesh border is estimated to be 120 km and along the curvatures it is approximately 220 km, extending over 50–70 km inside the mainland. At present the Indian Sundarbans covers an area of about 4,266.6 km^2. The tidal length in the Matla estuary along the sides of Gosaba Island is 85 km from the Bay of Bengal with spring tide range being 4.87 m and neap tide 2.13 m on an average.

The 'Dampier-Hodges Line' named after the Surveyor-Generals of India, Mr. Willam Dampier and surveyor Lt. Alexander Hodges, marks the northern limit of the Sundarbans, running in a slightly zigzag pattern from Basirhat in the north-east to Kulpi along the Hugli river in the west (Fig. 3.1). Politically the Sundarbans in the district of North-24 Parganas falls within the six Community Development (C.D.) Blocks of Minakhan, Haroa, Sandeshkhali-I and II, Hasnabad and Hingalganj while in South 24 Parganas it extends over thirteen C.D. Blocks of Gosaba, Basanti, Canning-I, Canning-II, Joynagar-I, Joynagar-II, Kultuli, Patharpratima, Namkhana, Sagar, Kakdwip and Mathurapur-II.

Our study area is the inhabited part of the Gosaba Island (Fig. 3.2). The Gosaba C.D. Block consists of 51 mouzas (the smallest revenue unit of administration) with

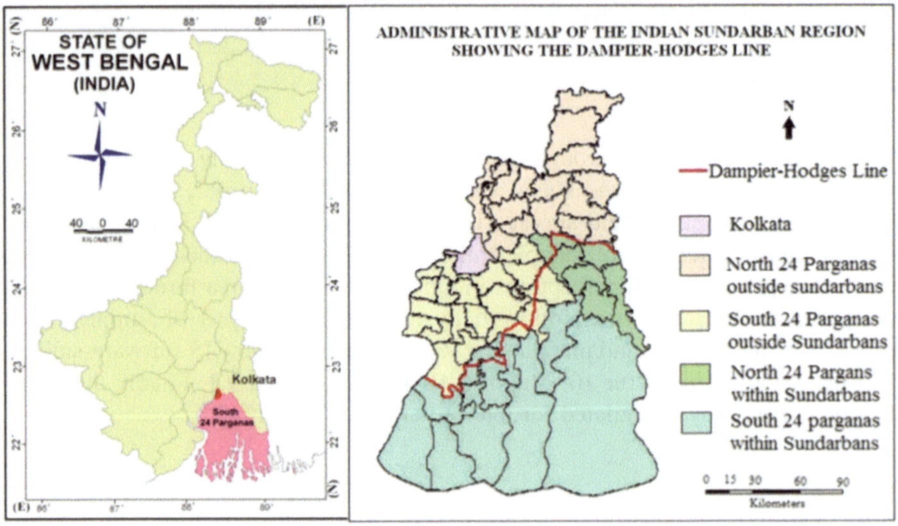

Fig. 3.1 Location of the Indian Sundarban Region, with respect to the state of West Bengal and Kolkata City

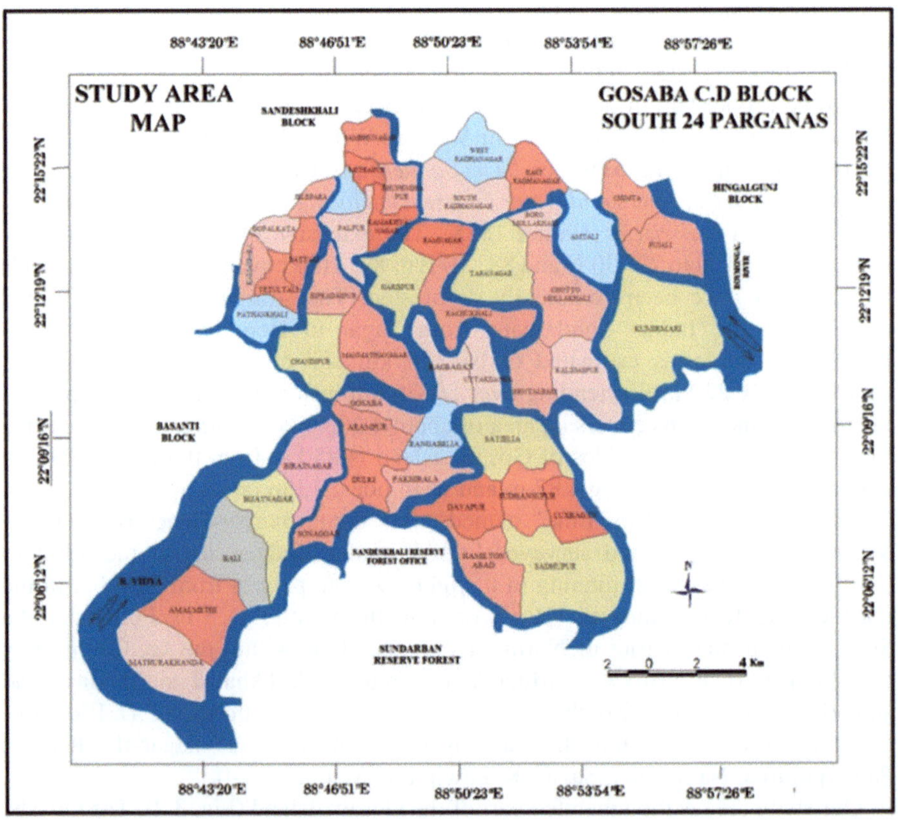

Fig. 3.2 Gosaba C.D. Block showing the villages under study

an average area of 5.6 ha. According to the District Statistical Handbook, Government of West Bengal, (Anon 2009), the decadal population growth rate was 11.2 % (1991–2001) with a population density of 799 persons/km^2. There were a total of 222,822 people in the Gosaba C.D. Block living in 44,478 households. Of the 16,896 cultivator population 2.44 % were leased cultivators, 7.1 % small farmers, 20.96 % sharecroppers and 40.53 % marginal farmers. As per the estimates about 44 % of the population live below the poverty line, 60 % of the households do not have access to clean drinking water, 87 % people suffer from food shortages, about 47.55 % households own no land, half (52.27 %) of the workers in agriculture and fishing are daily wage labourers and only 30 % of the families live in *pucca* (concrete and cement structures) or partially *pucca* houses (data from direct correspondence with Dr. Kalyan Rudra, Kolkata, January 2013).

3.5 Climatic Condition and Hazards (Cyclone)

The average monthly temperature is around 35–26 °C (April to June) and 24–14 °C (December to January). The average annual rainfall range is 1,600–2,000 mm, while average relative humidity is about 75–85 %. The summer (pre-monsoon) extends from the middle of March to mid-June, and the winter (post-monsoon) from mid-November to February. Rough weather with frequent cyclonic depressions occurs during mid-March to mid-September. The monsoon usually sets in around the middle of June and lasts up to the middle of October. But over the last decade it has been noticed that monsoons are delayed up to the end of July when high intensity rainfall occurs followed by long dry spells. The frequencies of pre-monsoon cyclones too have markedly increased. The most notable among them was the *Aila*, which had accompanied the high tides of the season leading to disaster (Table 3.1). The strong high sea-waves not only overtopped the embankments and flooded the islands but also washed away life and property. The ill-effects

Table 3.1 Damages caused by the Aila

Parameters	Values
Number of villages affected	4,249
Size of affected population	2,562,442
Number of people missing	8,000
Number of deaths	Official-70; Unofficial-300
Length of embankment breached	400 km
Number of cattle lost	212,851,212,851
Total area of agricultural land affected	125,872 ha
Estimated financial loss in agriculture	Rs. 337 crore
Number of houses fully damaged	194,390
Number of houses partially damaged	194,701
Total loss	Rs. 1,495.63 crore

Source: Rudra (2010), Aila: Unpublished Records of the Govt. of West Bengal. A South Asian Journal on Forced Migration, MCRG, Kolkata. pp 86–93

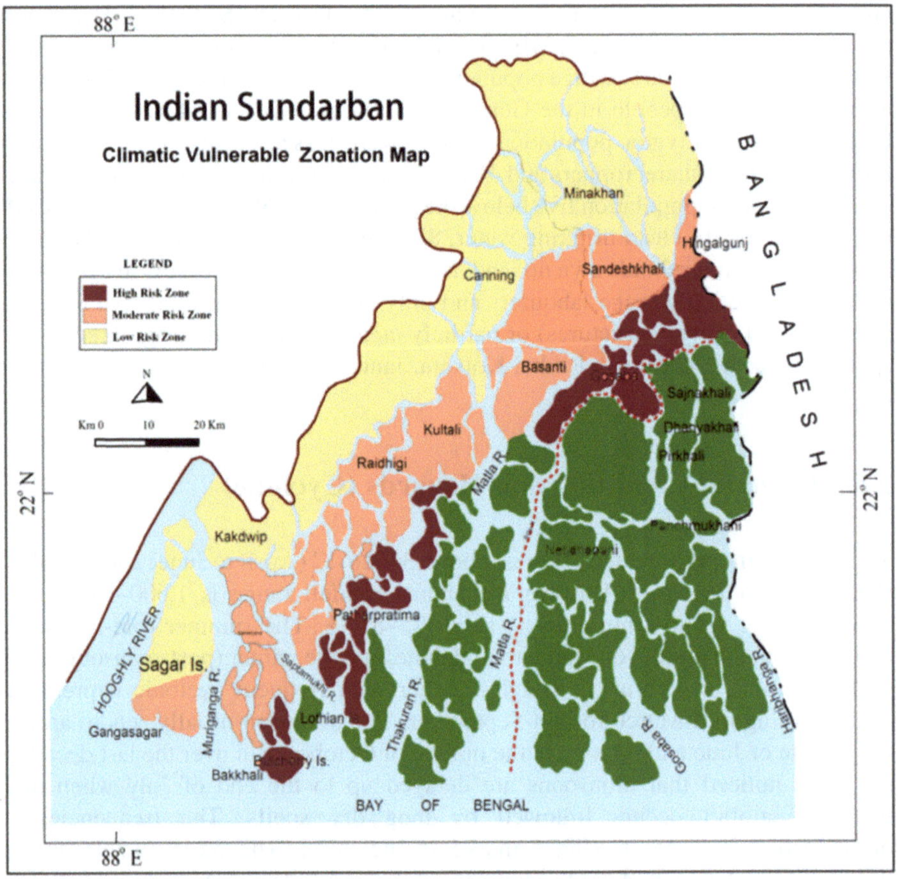

Fig. 3.3 Climatic Vulnerable Zonation Map of the Indian Sundarban Region. *Data Source*: India Meteorological Department Records

of *Aila* are still continuing in many areas of the Sundarbans. The area under study has been designated as a High Risk Zone according to India Meteorological Department (IMD) climatic hazard and disaster study parameters due to frequent hit by cyclones, flooding, embankment breaching, storm surges along with high density of population (Fig. 3.3).

3.6 Soil Characteristics

The soils of this area are mostly of heavy texture (silty clay to clay loam) hence have low hydraulic conductivity and often causes water logging that destroys the crop especially in the summer growing season. As observed from the field sample studies, the soils on an average contain 25 % sand, 45 % silt and 30 % clay.

The soil of Sundarbans are mostly fine brown to grayish black silts and dark clays and swampy soils at the sea-face. The soils are in general rich in potassium (K) and have moderate amount of available phosphorus (P) but low in nitrogen (N). The organic carbon content of the study area ranges from 0.73 to 0.78 is comparatively low resulting in slightly higher pH values (Soil Laboratory, Tagore Society for Rural Development TSRD, Rangabelia,). This is accompanied by occurrence of high amount of water soluble salts in the soil as evident from the electrical conductivity (EC) values-Gosaba: 4 and Lahiripur: 10.08 mmhos/cm (Sarkar et al. 1999a, b). The high soil salinization gives a whitish colour to the top-soil. The salinity and acidity of the soil influences the chemical transformation of nutrients and their availability to plants. Most mangrove soils are well buffered, having a pH in the range of 5–7. There is heavy siltation in the Sundarbans and the sediment trapping is aided by pneumatophores and dense mangrove outgrowths. The forested areas on the contrary have much higher available nitrogen and phosphorus content in the soil and high organic matter content, due to the poor decomposability of plant matter which helps in maintaining higher range of water holding capacity and thus lesser concentrations of soluble salts.

The WWF reports (Danda 2010; Danda et al. 2011) states that, the rate of groundwater flow across the southern boundary of Sundarban to be around 68 MCM/year. Generally groundwater from 160 to 400 m below ground level can been tapped in this region but are sometimes saline. This water is slightly alkaline with pH ranging between 7.8 and 8.2. It is further reported that the sub-ground sweet water storage at Basanti, Joytisrampur, Gosaba are far below than at Canning, Nimpit and Mandirbazar, and the subsurface saline water band is also much thinner (25 m in Canning compared to 150 m at Joytisrampur). They have also referred to the fact that there are three aquifer zones, the shallower ones occur at 60 m below ground level which is mostly brackish; the second band is at 70–160 m below ground level and also saline while the third aquifer zone contains sweet water found at 160–400 m below ground level. A layer of clay varying in thickness between 4 m in Gangasagar to 120 m at Kultuli separates the freshwater group of aquifers from the saline water aquifers. Thus irrigation through shallow pumps is not at all suitable in Gosaba and the adjacent islands.

3.7 Landuse Changes After *Aila*

The Figs. 3.4 and 3.5 have been prepared from LANDSAT 7 Satellite Image dated January, 2009 and 2010. From the supervised maximum likelihood classification it is observed that most of the agricultural lands lay unproductive and fallow during the winter season due to fresh water scarcity and salt encrustations. The dry season crop needs more irrigation and fertile soil for cultivation. Only a brief summer crop was possible in few plots that too with insufficient yield in 2010. The north-western part of the area is at a higher elevation, so the damage was much less (Pathankhali and Manmathanagar area) while maximum damage is observed in the eastern side in Hetalbari and Chotto Mollahkhali mouzas.

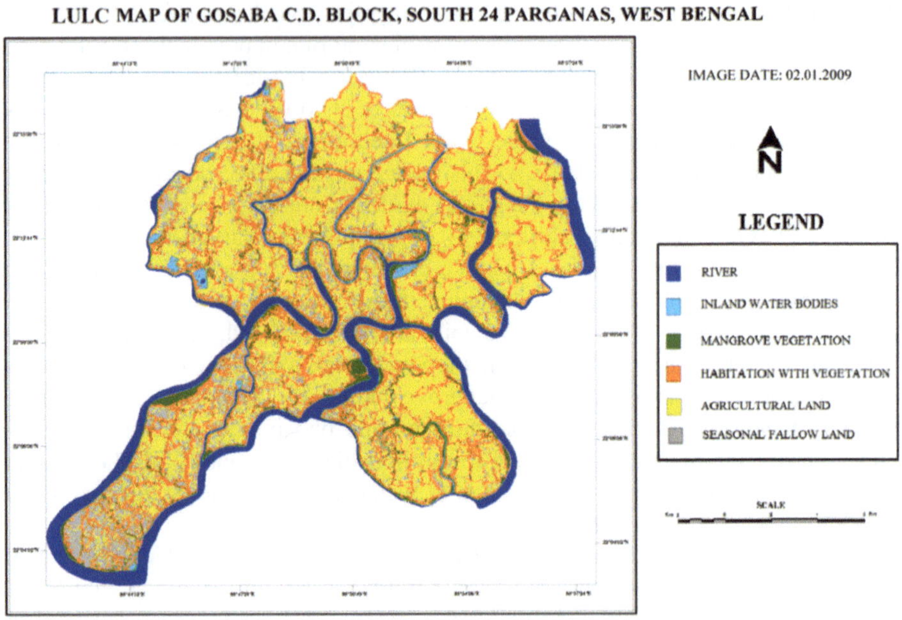

Fig. 3.4 Land-use and Land-cover (LULC) Map of Gosaba Island, in 2009 before Aila showing more area covered with agricultural land

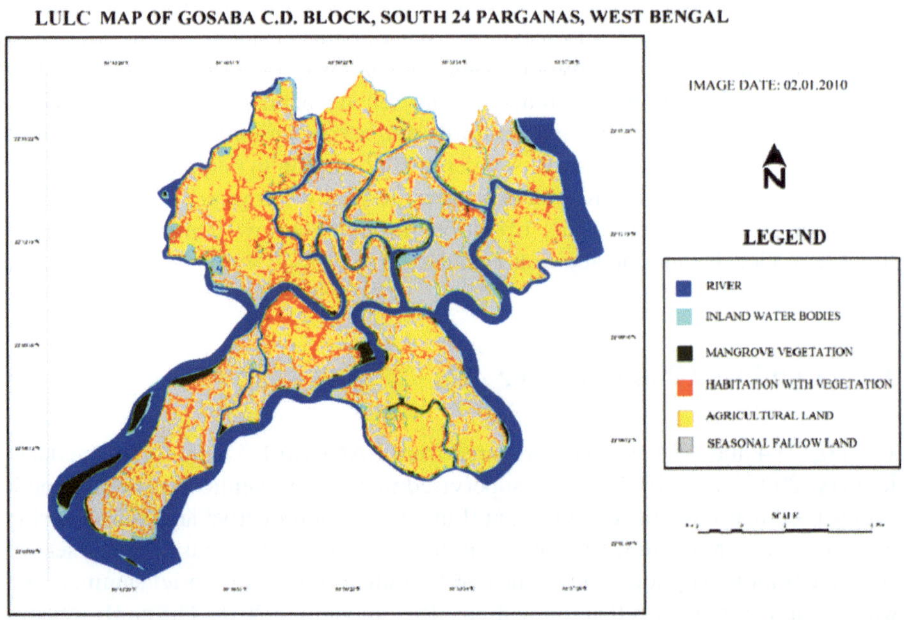

Fig. 3.5 Land-use and Land-cover Map of Gosaba Island, in 2010 after Aila showing major areas lying unproductive in the winter season

3.8 Impact of Salinity and pH on Agriculture of Gosaba Island

Soil salinity reflects the geophysical features of the ecosystem. It is also an indicator of crop productivity, dilution caused by surface run-off, stream discharge, barrage discharge and other anthropogenic activities. The pH of a soil significantly affects plant growth, primarily due to the toxicity from rising concentrations in availability of both essential and non-essential elements such as phosphorus (P) and aluminium (Al). As a result, all the elements depending on water availability are affected. The scarcity of the Ganges fresh-water flow too is a challenge for coastal food security, mangrove wetland ecosystems protection and further improvement of coastal saline environment in the Sundarbans.

A salinity level of creek and tidal river water is expected to raise the salinity of cultivated areas over which water overtops or if irrigated. As per our field data, it has been estimated from the water samples, that the pH level of the Matla estuary at Canning is 7.6–8.1 while salinity average is 25 ‰ at 25 °C. River Bidya has salinity of 20 ‰ and Pakhirala 24 ‰ with pH about 7.7 in both. The creeks around Gosaba town, records an average salinity of 24 ‰ in pre-monsoon, 16 ‰ in monsoons and 20 ‰ in post monsoon seasons. Salinity has been measured in ‰ also referred to as particles per thousand (ppt) or grams per kilogram. Salinity level remains least during the post-monsoon season due to huge fresh water inflow and rainfall that washes away the soluble salts whereas the maximum concentration is found in summers when high evapotranspiration draws out the soil moisture through capillary action, leaving the salt deposits on the top soil known as salt encrustations (Nath and De 1999) (Fig. 3.6).

As water above 2 ‰ salinity is unsuitable for irrigation hence lack of fresh water supply hinders crop productivity. From our field observations, it can be deduced that soil salinity increased with depth, while organic carbon and pH decreased with depth. Soil salinity decreased with increasing distance from the tidal coast but no

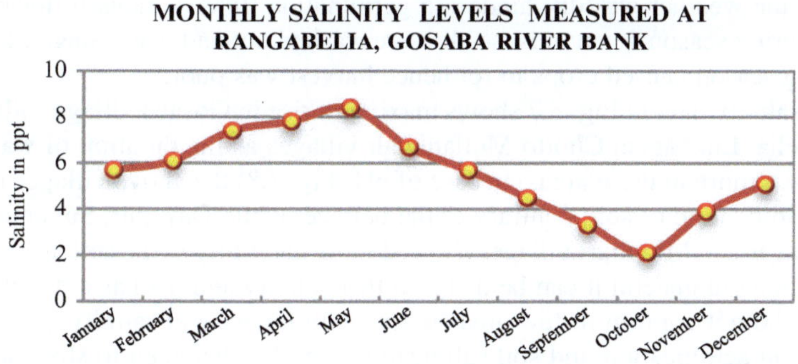

Fig. 3.6 Average monthly changes in the salinity levels in soil samples, calculated over a year at Rangabelia 2011 (Salinity at river banks are much higher than in agricultural fields). *Data Source*: Primary field records

such trend was noticed in soil pH. Frequency of tidal inundation too seemed to affect soil salinity. An increase of soil salinization, organic carbon and pH due to anthropogenic activities in this part of the Sundarbans was identified, which if continued may decrease the potential of Sundarban soil as a carbon sink and make the soil highly unproductive (Banerjee et al. 2012).

Extensive areas of Gosaba C.D. Block were marooned with saline water during *Aila* and the water stagnated over 2 months in most low lying fields. Our study since 2010 of the Gosaba island in the pre- and post-Aila scenario shows that even though alkalinity and acidic effects have been variable in these two locations, salinity have markedly increased from 0.1 to 5.25 ‰. This has rendered the soil unproductive. According to the local people the only effective measure known was washing off the area by three successive years of monsoon precipitation which can help to regain soil fertility. The water stagnated for about three months on the agricultural plots after Aila. Crop production was not possible in the *kharif* (summer) season of 2009. The paddy production in 2010 was almost negligible. In 2010 and 2011 due to variability of rainfall and poor soil quality few crops could be grown. New seeds, fertilizers, irrigation were used but eventually most failed and production was lower than expected. For example the *lunishree* variety of salt-tolerant paddy was introduced but this too failed to give sufficient yield. Untimely rains in 2012 made production even more difficult. By this time the infiltrated soluble salts started rising up the soil capillary and subsequent evaporation formed salt encrustations on the cultivable lands transforming them into barren infertile fallows. Thus the concept of top soil salt cleansing by washing out with fresh water in three successive years was nullified. More so with the use of HYV seeds, the fertilizer and irrigation needs increased. Most of the chemical fertilizers had negative effects on the soil fertility. Potash and pesticides were even used at very high costs but the effects were detrimental. Fresh water supply was drastically reduced by salt water ingression in ponds and reservoirs. The farmers could not afford deep tube-wells, diesel mechanized farm implements and hence could not even grow crops for family sustenance. The timing of onset and withdrawal was delayed. In the first few years after *Aila* farmers saw that during sowing period when water was most needed, the weather was dry hindering germination. Intense rains followed in the pre-maturing season when most seeds were washed out and water stagnation in the maturing season caused crops to rot hence harvest was poor.

The salinity data in Fig. 3.7 shows maximum rise in Gosaba village followed by Rangabelia, Luxbagan, Chotto Mollahkhali villages as the duration of water stagnation was more in these areas. In case of pH (Fig. 3.8) the above villages recorded more acidification of soil, contrary to the believe, while Dayapur, Sukumari, West Radhanagar and Mitrabari villages showed more alkalinity in the soil after the *Aila*. Even though in general it can be deduced that salinity and pH has a direct relation but this did not confirm in few areas where with increase in salinity, pH dropped. The rise in acidification and soil salinization may be attributed to slow decomposition of the biomass trapped in waterlogged conditions within the basin-like agricultural plots for months after the *Aila*.

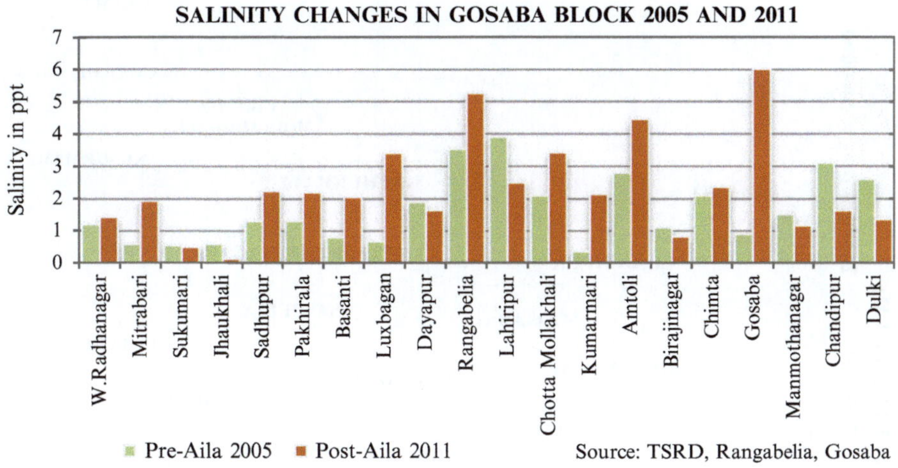

Fig. 3.7 Salinity changes of different villages in Gosaba Island

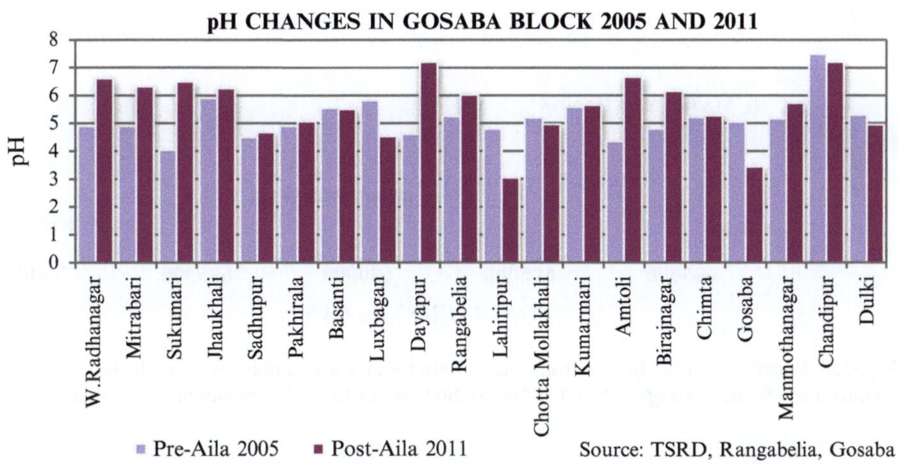

Fig. 3.8 pH changes of different villages in Gosaba Island

On the basis of the primary and secondary data analysis it is evident that salinity did increase after *Aila* in the region. The changes altered basic soil characteristics related to aeration, temperature, moisture and the organisms that live in the soil. Symptoms of excess chloride on crops include burning and firing of leaf tips or margins, bronzing, premature yellowing, abscission of leaves and, less frequently, chlorosis. Smaller leaves and slower growth also are typical. Symptoms of excess sodium include necrotic areas on the tips, margins, or interveinal area (Rhoades 1995).

A study had been done to record the salinity and pH in soil samples of the villages in the study area in winter season (Fig. 3.9). The isohaline maps (Fig. 3.10) on the sample villages shows the salinity distribution of the study area. Before Aila

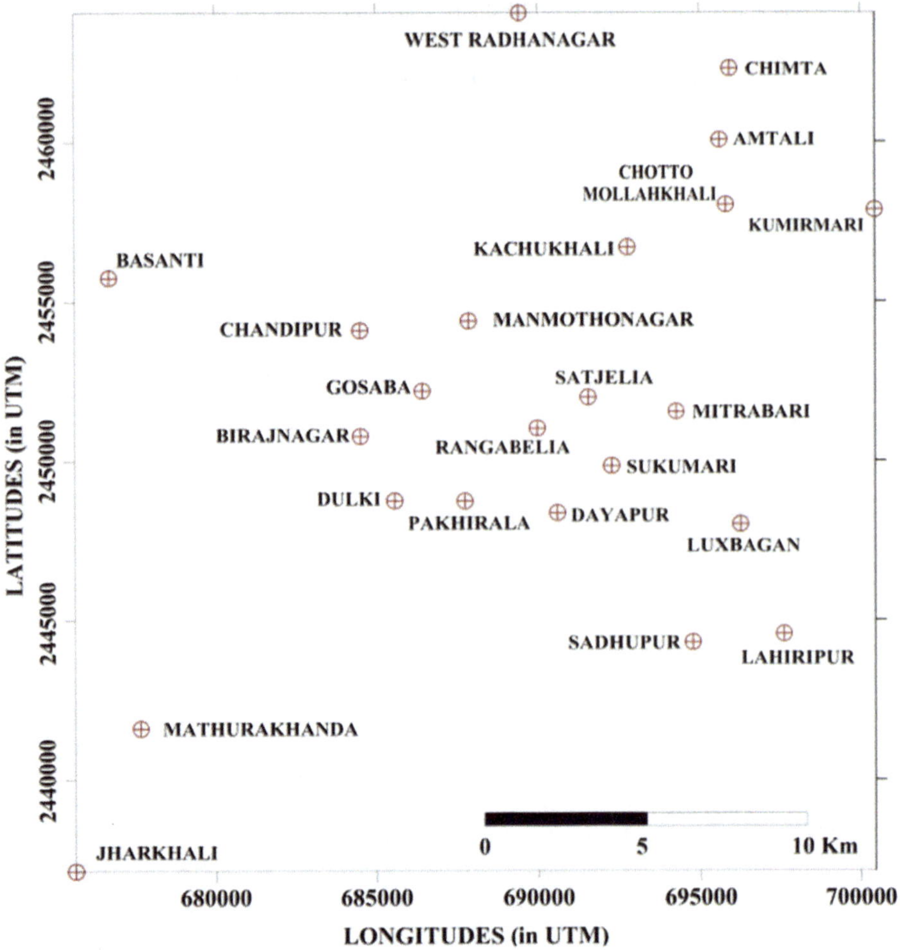

Fig. 3.9 Location of the villages from where Salinity and pH samples were collected, (Gosaba Island). Data Source of Figs. 1.9–1.11: Tagore Society for Rural Development, Rangabelia

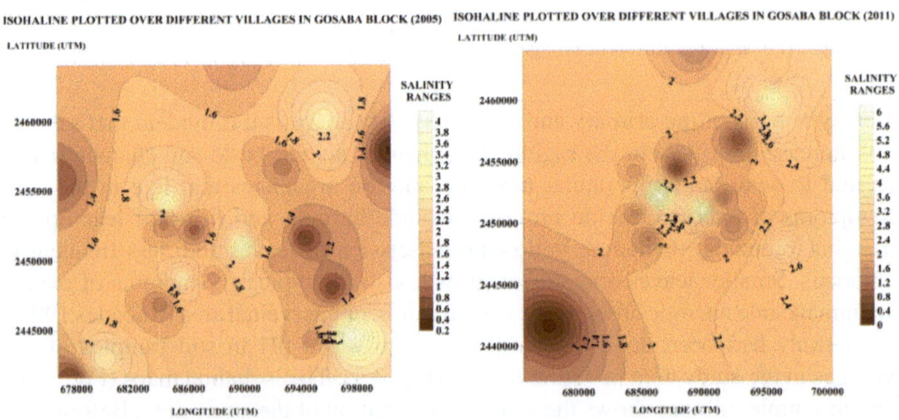

Fig. 3.10 Salinity Distribution Map, over villages around Gosaba (in 2005 and 2011)

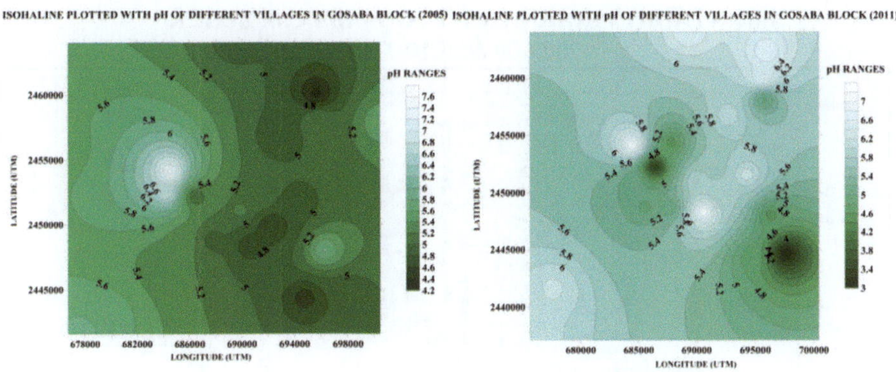

Fig. 3.11 pH Distribution Map, over villages around Gosaba (in 2005 and 2011)

Table 3.2 Salinity tolerance level of the commonly grown crops in the study area

Field crops	Productivity rate (%) as per salinity (values in ppt or ‰) 100 %	50 %	0 %	Relative crop salinity tolerance rating
Turnip (*Brassica rapa*)	0.603	4.355	8.04	Sensitive
Radish (*Raphanus sativus*)	0.804	3.35	5.963	Sensitive
Potato (*Solanum tuberosum*)	1.139	3.953	6.7	Moderately sensitive
Sugarcane (*Saccharum officinarum*)	1.139	6.7	12.73	Moderately sensitive
Cabbage (*Brassica oleracea capitata*)	1.206	4.69	8.04	Moderately sensitive
Spinach (*Spinacia oleracea*)	1.34	5.762	10.05	Moderately sensitive
Cucumber (*Cucumis sativus*)	1.675	4.221	6.7	Moderately sensitive
Tomato (*Lycopersicon esculentum*)	1.675	5.092	8.71	Moderately sensitive
Rice (paddy) (*Oriza sativa*)	2.01	4.824	7.37	Moderately tolerant

Source: Modified from FAO Corporate Document Repository (1994)

the average salinity was 1.4–1.6 ‰ which increased to 2–2.6 ‰. This shows that high salinity occurs in the central part of Gosaba Island and around the Choto Mollakhali Island but major part of the Sundarbans were under moderate salinity zone. The pH maps (Fig. 3.11) shows that the maximum area was within the average range of 4.5 which changed to 5.5.

All plants do not respond to salinity in a similar manner; some crops can produce acceptable yields at much greater soil salinity than others. This is because some are better able to make the needed osmotic adjustments enabling them to extract more water from a saline soil. The ability of the crop to adjust to salinity is extremely useful. In areas where a build-up of soil salinity cannot be controlled at an acceptable concentration for the crop being grown, an alternative crop can be selected that is both more tolerant of the expected soil salinity and can produce economical yields. Table 3.2 shows that the crops that originally grew in these areas due to conditions being more than their tolerance level, has stopped yielding. Only certain varieties of paddy have still shown more than 80 % production where salinity has

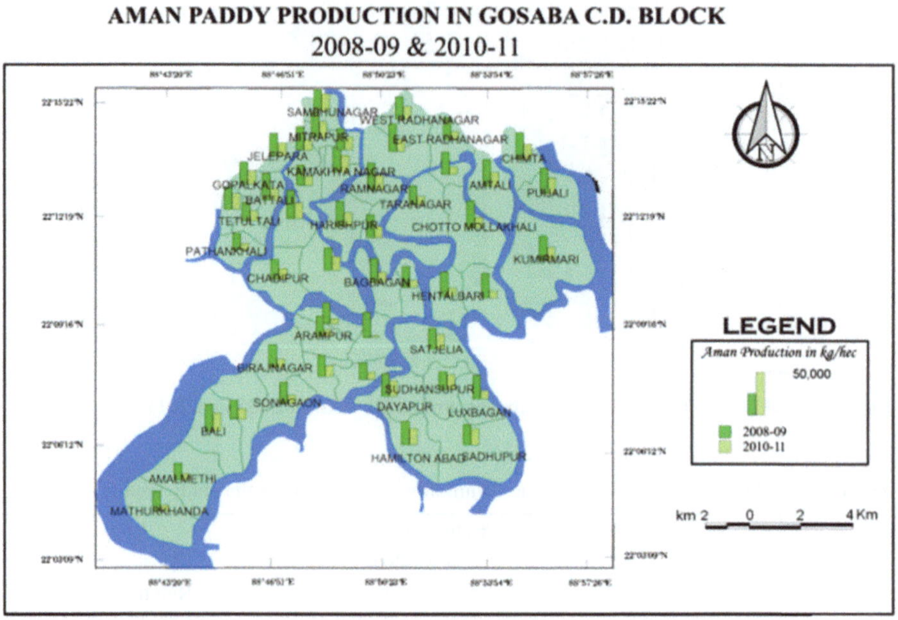

Fig. 3.12 A comparative representation of *Aman* paddy production in Gosaba Island before and after Aila. Data Source: Agriculture Development Office (A.D.O.), Gosaba, W.B

not risen above 2 ppt. Tomato and cucumber has shown a poor crop in 2012 while green vegetables including cabbage and spinach are difficult to be grown.

The agricultural system and pattern markedly changed so as to adapt with the changing climatic regime. Warming of the weather induces pests and diseases which also destroys the crops. Due to high soil salinity and lack of irrigation in dry months the land is left fallow except for a single rice crop in *kharif* season. 80 % of the agriculture land is under monoculture, while only in 20 % of the plots are used for crop rotation with few vegetables and legumes in seasons where irrigation is available. This causes perceptible changes in the land use of the region. Agricultural and forest productivity is fast diminishing due to salinity rise and conversion of agricultural fields to aquaculture ponds (Hoque et al. 2006). The *bheries* (aquaculture pond) that have turned unproductive are being converted to brick fields to increase the profits. This has a large impact on the ecosystem of the region. Brick kilns here are known to degrade the soil which can only be used for construction or left barren in future. This markedly alters the geomorphology of the region.

Aman (monsoon) paddy production shows a fall in production from 28,000 kg/ha in 2008–2009 to 14, 000 kg/ha in 2010–2011 in most of the villages. *Boro* (winter) paddy showed even greater fall from 34,000 kg/ha in 2008–2009 to 20,000 kg/ha in 2011–2012. Negligible production occurred during 2010–2011 hence was not recorded by the Agricultural Development Office (A.D.O.), Gosaba. In Chotto Mollahkali *aman* yield fell from 34,000 kg/ha to 12,000 kg/ha. while in Sadhupur *boro* yield fell from 37,000 kg/ha to 19,000 kg/ha in the post-*Aila* period. This change has been shown through bar diagrams in Figs. 3.12 and 3.13.

BORO PADDY PRODUCTION IN GOSABA C.D. BLOCK
2008-09 & 2012-13

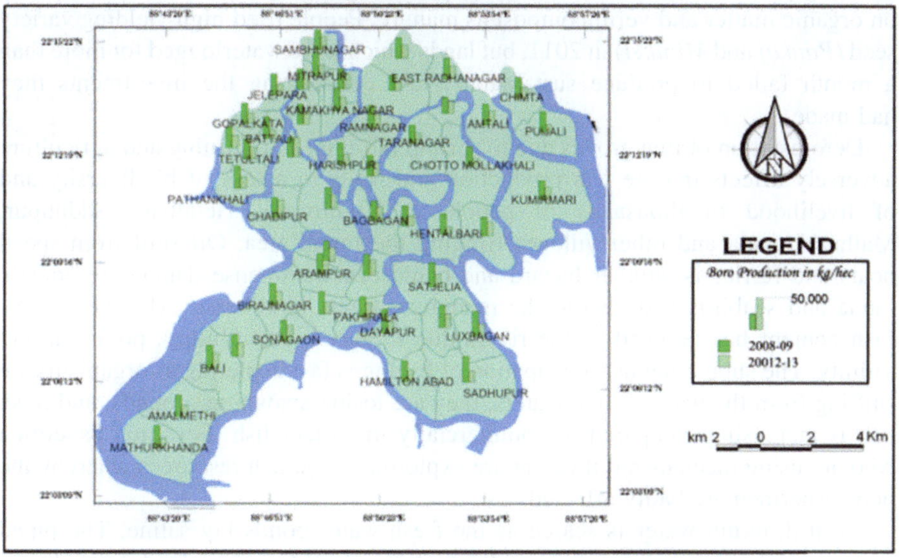

Fig. 3.13 A comparative representation of *Boro* paddy production in Gosaba Island before and after Aila. Data Source: A.D.O., Gosaba, W.B

3.9 Effect on Human Livelihood

Since a large portion of the population is still dependent on monsoon rains, there is a considerable yearly fluctuation in crop yields and productivity. Variations in weather conditions in the form of local storms, cyclones, floods and dry spells often leads to colossal damages to food production systems. This directly increases the market price of food products in the local markets. Even after three consecutive rainy seasons, salinity remains alarmingly high in the soil hence agriculture, the major occupation of the region, is far from prospering.

Before Aila the landless wage labourers' earnings were Rs. 100–120 per day, but after Aila it has increased to a minimum of Rs. 200 in addition to food per day. This is due to extensive migration of the workers in search of other jobs elsewhere. A tractor charges Rs. 300 to till 1.67 acres of land in addition to the rising diesel prices. It was observed that 15–20 years ago no fertilizer was needed. Now not only are the fertilizer costs increasing Rs. 10–15 yearly but also becoming ineffective. In addition to these, the farmers have to employ labourers at the time of sowing and harvesting at the rate of Rs. 120 per labourer per day. Draft animals perished in the deadly storm hence most of the labour has to be hired at steep rates. This has increased the capital cost of farming. With the unsatisfactory yields the minimum costs has not being recovered by the cultivators till today. Only 72 % of the households have only 1 acre of cultivable land hence cannot afford mechanized implements, costly pesticides fertilizers and irrigation by shallow pumps thus reducing productivity. There is no facility for crop storage. Farmers can only rely on organic matter and vermicompost as manure. People tried high yielding variety seed (*Pankaj* and *Miniket*) in 2011, but lands which were waterlogged for more than a month failed to produce sustainable yield considering the investments they had made.

Deforestation of mangroves due to shrimp farming, salt farming and agriculture adversely affects marine fish production and leads to a loss of biodiversity and of livelihood to thousands of people in Kumirmari, Hetalbari, Sadhupur, Mathurakhanda and other villages fringing the forest area. Oil spill from speed boats and ferries is another hazard and could cause immense damage to aquatic fauna and seabirds and also to the mangrove forest biodiversity (Blower 1985). Fish content has reduced in the rivers due to excessive turbidity, pollution and salinity. The supplementary occupation of fishing is fast depleting as organisms are shifting from the warm saline waters, they are losing spawning grounds, and trash fish is increasing compared to commercially important fish. Few rich fishermen who are using mechanized trawlers are exploiting large fish resources whereby the poor fishermen are badly affected.

Even drinking water is scarce as the fresh-water ponds lay saline. The piped water from the Sonagao reservoir on the eastern side of Gosaba Island is not always clean and hygienic. There were attempts to desalinate the ponds soon after the broken and collapsed embankments were repaired but were not successful due to lack of funds. Immediately after the cyclone the government declared Rs. 2,500 and Rs. 10,000 as compensation for partially and fully damaged mud houses respectively but only a few received it. The poorer section did not get any compensation when they most needed it, due to political reasons and mal-practices. All these have left the working population with little choice than to migrate to other areas. In Satjelia, Dayapur villages of the Gosaba Island it was found that almost 75 % of the male working population had left their homes in search of employment to Kolkata city, Tamil Nadu, Gujarat, Kerala and other places.

3.10 Conclusion

The yearly natural calamity, variability of monsoon rainfall and its impacts are new threats to food security and biodiversity in the region. 60 years of development and 230 years of settlement was shattered in just few hours in the *Aila* cyclone in 2009. This majorly affected the agricultural sector of this region. The soil turned too saline to cultivate, crop productivity fell, rapid bank erosion, embankment breaching lead to a struggle for existence for the inhabitants. Local farmers experiencing not only a loss in cash crops but also subsistence crops are now threatened. Therefore, the increased salinity and alkalinity has not only damaged crop productivity and agricultural pattern but also changed the cultural landscape in this region. There is a constant threat from the cause and effects of climate changes and is affecting the food security and livelihood in Sundarbans.

Hence an adaptation strategy needs to be enhanced for wider application and identification of appropriate agronomic management practices. Capacity building strategies that strengthen developing countries in coping with extreme weather events, agricultural sustainability, water conservation and hazard insurance must be focused at the earliest (Shukla et al. 2013). The effect of other land use and land cover change driving forces need to be included to have a realistic estimate of the impacts. There is a need to reduce our reliance on irrigation, hazardous chemical fertilizers and pesticides that turn the soil saline. An early warning system, development of salt tolerant and pest resistant crop that can adjust to the changing climatic regime and biotechnological advancements are essential. Paddy being the most resistant crop, various salt tolerant new varieties can be cultivated like *Hogla, Hamai, Kalomota, Katarangi, SR26B, Sabita* and *Lunishree*. Introduction of no-till farming practices reduces soil erosion and methane emissions. Long fallow periods particularly during hot summers must be avoided when rapid secondary soil salinization from high water tables can occur. Rain water harvesting, proper soil management techniques and adequate drainage facility are equally important as long-term salinity control which is not otherwise possible.

Thus from the observations on soil salinity and crop productivity problems and changes in livelihood pattern, it can be only suggested that superficial grants and aids are no solution to mitigate the miseries of the affected population who, at times, do not even receive the basic needs in life. As natural calamities are inevitable, these problems tend to be recurrent, although with much lesser intensities. The welfare organizations need to concentrate more on the adaptive side of the hazards to be able to effectively cope with disasters. This can be done by developing flexibility of the cultivators to alter their cropping pattern and offer a wide variety of crops at low cost to choose the appropriate in times of need and earn profits, thus ensuring better living conditions to the residents and over-all development of this region.

Acknowledgments We are greatly obliged to our guide and Associate Professor Dr. Lakshminarayan Satpati (University of Calcutta) for his constant support and encouragement. We are extremely grateful to Dr. Kalyan Rudra (river scientist), and acknowledge the help of Dr. Tushar Kanjilal and the officials at Tagore Society for Rural Development, Rangabelia and at the Agricultural Development Office, Gosaba, South 24 Parganas District, West Bengal. We also thank Anish Bhattacharyya and fellow research scholars in the Department of Geography, University of Calcutta, India.

References

Anon (2009) District Statistical Handbook, South 24 Parganas, Bureau of Economics and Statistics, Government of West Bengal

Banerjee K, Roy Chowdhury M et al (2012) Influence of anthropogenic and natural factors on the mangrove soil of Indian Sundarbans wetland. Arch Environ Sci 6:80–91

Blower J (1985) Sundarbans forest inventory project, Bangladesh. Wildlife conservation in the Sundarbans. Project report 151. Overseas Development Administration, Land Resources Development Centre, Surbition, UK, p 39

Danda AA (2010) Sundarbans: future imperfect climate adaptation Report. World Wide Fund for Nature, New Delhi

Danda AA, Sriskanthan G, Ghosh A, Bandyopadhyay J, Hazra S (2011) Indian Sundarbans delta: a vision. World Wide Fund for Nature, New Delhi

FAO Corporate Document Repository (1994) Water quality for agriculture in FAO Irrigation and Drainage Papers. Version 29. Agriculture and consumer protection. Food and Agriculture Organization of the United Nations. http://www.fao.org/docrep/003/t0234e/T0234E03.htm, accessed on 25. 4.2013

Hoque MA et al (2006) Present status of salinity rise in Sundarbans area and its effect on Sundari (Heritiera fomes) species. Res J Agr Biol Sci 2(3):115–121, INSInet Publication, p 119

Hunter WW (1875) A Statistical account of Bengal, 1 (Districts of 24 Parganas and Sundarbans). Trubner and Co., London

Islam SN, Gnauck A (2008) Mangrove wetland ecosystems in Ganges-Brahmaputra delta in Bangladesh. Front Earth Sci China 2(4):439–448

Kalra N, Aggarwal PK, Chander S, Pathak H, Choudhury R, Chaudhary A, Sahgal M, Rai HK, Soni UA, Sharma A, Jolly M, Singh UK, Ahmed O, Hussain MZ (2003) Impacts of climate change on agriculture. In: Shukla PR, Sharma SK et al (eds) Climate change and India Vulnerability Assessment and Adaptation. University Press (India) Pvt. Ltd, Hyderabad, pp 193–223

Khan AA, Hoque MA (2002) Quaternary Paleogeography and Geohazard scenario of the Bengal delta of Bangladesh. Orient Geogr 46(2):1, www.ucl.ac.uk/lag/docstorage/Khan_Hoque2002.pdf Accessed 1 May 2013

Mitra S, Bhadwal S, Kelkar U, Rangan L (2010) Climate change vulnerability in Indian agriculture: a major hurdle toward food security. In: Singh RB (ed) Natural hazards and disaster management-vulnerability and mitigation, Rawat Publications, New Delhi, pp 274–284

Nath D, De DK (1999) Water and soil characteristics of different estuaries in Sundarbans. In: Guha Bakshi DN, Sanyal P, Naskar KR (eds) Sundarbans Mangal. Naya Prokash, Kolkata, pp 128–130

Rhoades JD (1995) Overview: diagnosis of salinity problems and selection of control practices. In: Tanji KK (ed) Agricultural salinity assessment and management. Scientific Publishers, Jodhpur, pp 18–41

Rudra K (2010) Aila: unpublished records of the Government of West Bengal. A South Asian Journal on Forced Migration. MCRG, Kolkata, pp 86–93

Sanyal P (1999) Global warming in Sundarbans delta and Bengal Coast. In: Guha Bakshi DN, Sanyal P, Naskar KR (eds) Sundarbans Mangal. Naya Prokash, Kolkata, pp 111–112

Sarkar D, Sen N, Naskar K (1999a) Nutrient status of different Mangrove Zones of the Indian Sundarbans. In: Guha Bakshi DN, Sanyal P, Naskar KR (eds) Sundarbans Mangal. Naya Prokash, Kolkata, pp 159–160

Sarkar D, Chattopadhyay GN, Naskar K (1999b) Nature and properties of Coastal Saline Soils of Sundarbans with relation to Mangrove Vegetation. In: Guha Bakshi DN, Sanyal P, Naskar KR (eds) Sundarbans Mangal. Naya Prokash, Kolkata, pp 201–202

Shukla PR, Sharma SK, Garg A, Bhattacharya S, Ravindranath NH (2013) Climate change vulnerability assessment and adaptation: the context. In: Shukla PR, Sharma SK et al (eds) Climate change and India vulnerability assessment and adaptation. University Press (India) Pvt. Ltd, Hyderabad, p 4, 12–14

Chapter 4
Long Term Rainfall Trend Analysis (1871–2011) for Whole India

Sananda Kundu, Deepak Khare, Arun Mondal, and Prabhash Kumar Mishra

Abstract Climate change has aroused serious consciousness among human beings as it has a strong impact on different parameters like rainfall, temperature, evapotranspiration etc. Change in climatic parameters also affects the agriculture and water demand of an area. The changed pattern of rainfall leads to extreme conditions like flood, drought and cyclones which have increased in frequency in the last few decades making the rainfall trend analysis extremely important for India where a large part of the economy depends upon rain-fed agriculture. The trend of rainfall for 141 years of India was analyzed in the present study from 1871 to 2011. Detection of trend was done by analyzing 306 stations of India divided into seven regions of Homogeneous Indian Monsoon, Core-Monsoon India, North West India, West Central India, Central Northeast India, North East India and Peninsular India. Temporal as well as spatial rainfall variability was shown on monthly, seasonal and annual basis. The Mann–Kendall (MK) Test and Sen's slope was applied in the study. Mann–Whitney–Pettitt (MWP) test was used to give the break point in the series. Annually, 5 regions have decreasing trend

S. Kundu (✉) • D. Khare • A. Mondal
Department of Water Resources Development and Management,
Indian Institute of Technology Roorkee, Roorkee, India
e-mail: sanandakundu@gmail.com; kharefwt@gmail.com; arun.iirs@gmail.com

P.K. Mishra
Department of Water Resources Development and Management,
Indian Institute of Technology Roorkee, Roorkee, India

Scientist, Water Resources Systems Division, National Institute of Hydrology,
Roorkee, India
e-mail: erprabhash@gmail.com

M. Singh et al. (eds.), *Climate Change and Biodiversity: Proceedings of IGU Rohtak Conference, Vol. 1*, Advances in Geographical and Environmental Sciences,
DOI 10.1007/978-4-431-54838-6_4, © Springer Japan 2014

except for core-monsoon and north-east India. Monsoon season depicted decrease in the rainfall magnitude in most of the regions. This result is extremely significant as monsoon rainfall serves the major water demand for agriculture. Change Percentage for 141 years had shown rainfall variability throughout India with the highest increase in North-West India (5.14 %) and decrease in Core-monsoon India (−4.45 %) annually.

Keywords Change percentage • India • MK test • MWP test • Rainfall trend

4.1 Introduction

With the growing frequency of different extreme events like flood, drought, cyclones etc., there has been increasing concern regarding the adverse effects of climate change. Changing rainfall pattern that is occurring because of climatic variations have raised the concern about extreme events (IPCC 2007). Evidences of climate change are extremely prominent in various sectors and it has been found that rainfall variation may have a significant impact on the agricultural and water of Asia-Pacific (Cruz et al. 2007). Increasing rainfall trend was observed in New York of USA (Burns et al. 2007), Australia (Suppiah and Hennessy 1998) etc. while a decrease was found in Italy (Buffoni et al. 1999) and Kenya (Kipkorir 2002). There were greater variation in the spatial and temporal distribution of rainfall like in Spain (Rodrigo et al. 2000) and North America (Englehart and Douglas 2006). Rainfall distribution and variability is also related to the temperature pattern which was found to be quite varied in Italy (Brunetti et al. 2000), Western and Central Europe (Moberg and Jones 2005), South Africa (Kruger 2006), Iran (Masih et al. 2010) etc. All these changes in rainfall trend have initiated more research in this field to analyze the distribution.

Both parametric and non-parametric tests were used for the trend analysis, but non-parametric tests were considered as better as they can be used on independent data sets (Hamed and Rao 1998). Mann–Kendall tests are non-parametric tests used most frequently for the trend analysis (Yue et al. 2003; Singh et al. 2008; González et al. 2008).

In India monsoon rainfall forms an important part of the climatic characteristics. It supplies the majority of rainfall required for the agricultural sector. The south-west monsoon supplies rainfall in north India while the winter or north-east rainfall serves the south India. Variation in rainfall have resulted in increasing threat over the economy of the country as there is rising problem in the agricultural sectors and planners must allocate the water resources properly to avoid problem of water scarcity in the future. In the present study, rainfall trend analysis of the entire country was done with the data of 141 years from 1871 to 2011 with 7 major regions of India. Mann–Kendall test and Mann–Whitney–Pettitt test was applied to the rainfall data on monthly, annual and seasonal basis.

4.2 Study Area

Indian rainfall ranges from 160 to 1,800 mm/year all over the country varying both spatially and temporally. As India is a large country with different climatic condition, 7 regions were formed for whole India on the basis of meteorological characteristics. These regions are, Homogeneous Indian Monsoon, Core-Monsoon India, North West India, West Central India, Central Northeast India, North East India and Peninsular India (except the hilly regions in the north) as given in Fig. 4.1. The study area covers about 2,880,000 km^2 areas for whole India. The data were assessed by annual, seasonal and monthly basis where summer months extend from March to May, monsoon from June to September, post-monsoon from October to November and winter from December to February. A total of 306 stations located all over India were analyzed for the study.

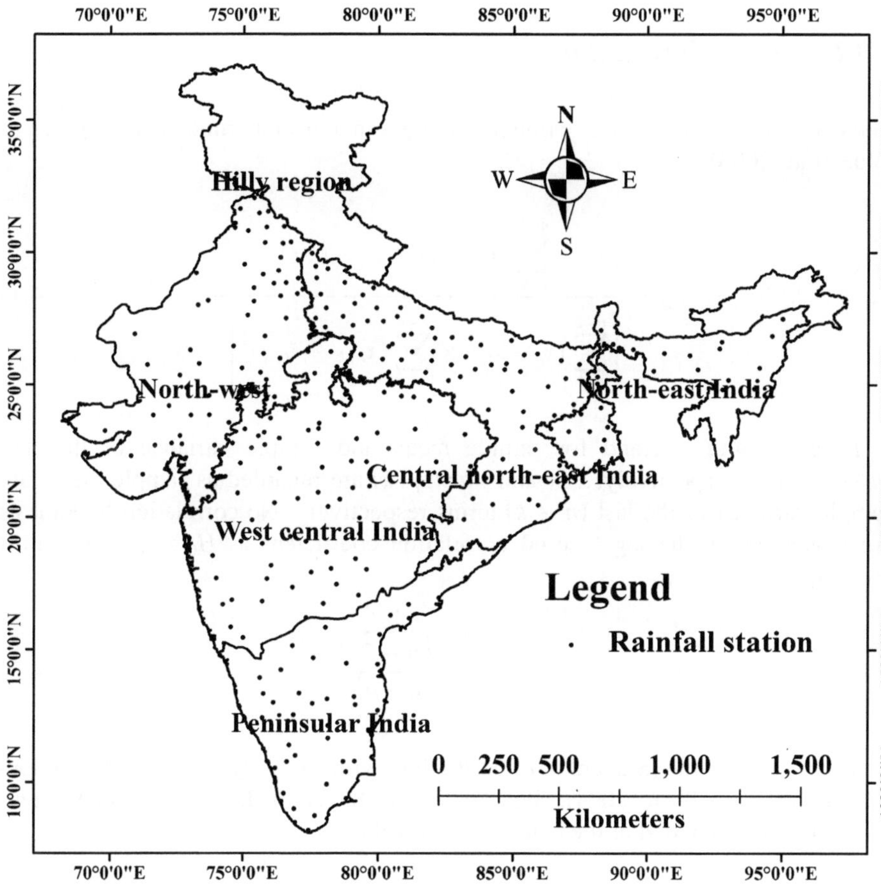

Fig. 4.1 Study area

4.3 Methodology

Monthly, seasonal and annual rainfall trend series was studied for 141 years to analyze the trend. Homogeneity test was applied on the series with Standard Normal Homogeneity Test (SNHT) at the 5 % significance level (Alexandersson 1986) and the observed series was found to be homogenous (Khaliq and Ouarda 2007). Serial correlation and pre-whitening method was applied on the series to remove the effect of correlation on MK Test (Storch 1993). MK test was then applied on the pre-whitened rainfall series (Mann 1945; Kendall 1975) with the Theil–Sen's estimator (Theil 1950; Sen 1968) to get the trend and magnitude of rainfall. Then Mann–Whitney–Pettitt (MWP) test was done to find the change point of the series (Pettitt 1979). At the end percentage of change was calculated (Yue and Hashino 2003).

4.3.1 *Serial Correlation*

The coefficient of serial correlation ρk for lag-k in a discrete time series is given as (Yue et al. 2003)

$$\rho_k = \frac{\sum_{t=1}^{n-k} (x_t - \overline{x}_t)(x_{t+k} - \overline{x}_{t+k})}{\left[\sum_{t=1}^{n-k} (x_t - \overline{x}_t)^2 \times \sum_{t=1}^{n-k} (x_{t+k} - \overline{x}_{t+k})^2\right]^{1/2}} \tag{4.1}$$

Here $\overline{x}_t$ and x_t stand for sample mean and sample variance of the first $(n - k)$ terms respectively, $\overline{x}_{t+k}$ and $Var\ (x_t{+}k)$ are regarded as sample mean and sample variance of the last $(n - k)$ terms respectively. No correlation hypothesis are examined by the lag-1 serial correlation coefficient as $H_0 : \rho_1 = 0$ against $H_1 : |\rho_1| > 0$

$$t = |\rho_1|\sqrt{\frac{n-2}{1-\rho_1^2}} \tag{4.2}$$

The t test is the Student's t-distribution with $(n - 2)$ degrees of freedom (Cunderlik and Burn 2004). For $|t| \geq t_{\alpha/2}$, the null hypothesis about zero correlation is discarded at the significance level α.

4.3.2 *Mann–Kendall Test and Theil–Sen's Estimator*

The MK statistic is given as

$$Z_c = \begin{cases} \frac{S-1}{\sqrt{Var(S)}} \ if,\ S > 0 \\ 0\ if,\ S = 0 \\ \frac{S+1}{\sqrt{Var(S)}} \ if,\ S < 0 \end{cases} \tag{4.3}$$

where,

$$S = \sum_{i=1}^{n-1} \sum_{j=i+1}^{n} \mathrm{sgn}(x_j - x_i) \tag{4.4}$$

Here x_j and x_i give the data values which are in sequence with n data, sgn (θ) is equivalent to $1, 0$ and -1 if θ is more than, equal to or less than 0 respectively. If Z_c is more than $Z_{\alpha/2}$ then the trend is regarded as significant and α is the level of significance (Xu et al. 2003).

$$\beta = median(X_i - X_j/i - j)\ \forall j < i, \tag{4.5}$$

where $1<j<i<n$ and β estimator represent the median of the entire data set (Xu et al. 2003).

4.3.3 *Mann–Whitney–Pettitt Method (MWP)*

The n length time series {X1, X2..., Xn} is taken. *t* represents the time of the most expected change point. Two samples {X1, X2,..., Xt} and {Xt+1, Xt+2,..., *X*n} can be attained by dividing the time series at *t* time. The *Ut* index is calculated in the following way:

$$U_t = \sum_{i=1}^{t} \sum_{j=t+1}^{n} \mathrm{sgn}(X_i - X_j) \tag{4.6}$$

where,

$$\mathrm{sgn}(x_j - x_i) = \left\{ \begin{array}{c} 1.......\ if\,(x_j - x_i) > 0 \\ 0.......\ if\,(x_j - x_i) = 0 \\ -1.......\ if\,(x_j - x_i) < 0 \end{array} \right\} \tag{4.7}$$

Constantly increasing value of |Ut| with no change point will be obtained by plotting Ut against t in a time series. But if there is a change point, then $|Ut|$ will increase up to the change point level and then will decrease. The significant change point t represents the point where value of $|Ut|$ is highest:

$$K_T = \max_{1 \leq t \leq T} |U_\tau| \tag{4.8}$$

The approximated significant probability $p(t)$ for a change point (Pettitt 1979) is:

$$p = 1 - \exp\left[\frac{-6K_T^2}{n^3 + n^2}\right] \tag{4.9}$$

When probability $p(t)$ overcomes (1–α), then the change point is significant statistically at time t with the significance level of α.

4.4 Results and Discussion

Homogeneity test was conducted for the series where T_0 at the 95 % level was found to be less than 9.468 (Table 4.3) (Khaliq and Ouarda 2007). Table 4.1 represents the serial correlation values for seven regions and gives the coefficients of monthly, seasonal and annual rainfall. The Lag-1 serial correlation was calculated to observe the presence of any positive or negative correlation and pre-whitening was applied to eliminate the effect. Maximum correlation was found in Peninsular India and North-West India (0.228).

The rate of change for 141 years for each month and the level of significance is given in Fig. 4.2. Positive and negative trends in the series were calculated from the MK test and some significant values were observed in different regions. The rainfall trend was significant in the month of April showing highest positive or increasing trend and decreasing trend was observed in July (<−15 %) so there was declining rainfall trend during the monsoon period. Significant negative rainfall trend was found in the month of July in the West Central India where the rate of change was −10 to −15 % over 141 years. In North-east India significant negative rainfall trend was noticed in the month of August (−10 to −15 %). The month of June also had a negative rate of change in the West Central India.

Apart from the monsoon months of June, July and August, decreasing trend was also noticed in December. Thus rainfall reduction in July–August or monsoon months indicates a problem for the agricultural economy. The north-west and peninsular India have mostly positive change rate in 141 years, unlike other regions particularly during July–August. The peninsular India experience the effect of both south-west and retreating monsoons which might have caused more rainfall here. However, during the time of retreating monsoon in Peninsular India, the rate of change was comparatively low (0–5 %). The North-west India also receives more rainfall.

Table 4.1 Serial correlation

Station name	Whole India	Homogeneous Indian monsoon	Core-monsoon India	North West India	West Central India	Central Northeast India	North East India	Peninsular India
Jan	0.142	0.082	0.114	0.127	0.062	−0.011	0.022	0.228
Feb	−0.012	−0.025	0.028	−0.059	0.051	−0.064	0.006	−0.027
Mar	−0.091	−0.08	−0.053	−0.029	−0.106	−0.102	−0.147	−0.049
Apr	−0.103	−0.097	−0.083	0.158	−0.122	−0.077	−0.05	0.022
May	−0.003	0.042	0.078	0.01	0.015	0.062	0.059	−0.066
June	0.004	0.073	0.091	0.173	0.042	−0.066	0.127	−0.127
July	−0.035	−0.114	−0.09	−0.092	−0.071	−0.12	−0.071	0.023
Aug	−0.093	−0.067	−0.054	−0.059	−0.034	−0.026	−0.037	−0.123
Sept	−0.039	0.035	0.065	−0.038	0.081	−0.127	0.013	−0.06
Oct	0.13	0.213	0.194	0.228	0.181	0.011	−0.055	0.045
Nov	−0.008	−0.034	−0.046	0.166	−0.055	−0.053	−0.121	0.088
Dec	0.055	0.025	0.065	−0.161	0.102	−0.045	−0.115	−0.033
Annual	−0.005	−0.004	0.028	−0.039	0.051	−0.062	0.063	0.042
Pre-monsoon	−0.076	−0.006	0.041	0.029	0.017	−0.079	−0.038	0.021
Monsoon	−0.092	−0.091	−0.055	−0.059	−0.051	−0.015	−0.001	−0.155
Post-monsoon	0.118	0.203	0.152	0.212	0.173	0.021	−0.075	0.005
Winter	0.02	0.044	0.078	−0.049	0.119	−0.04	−0.034	0.031

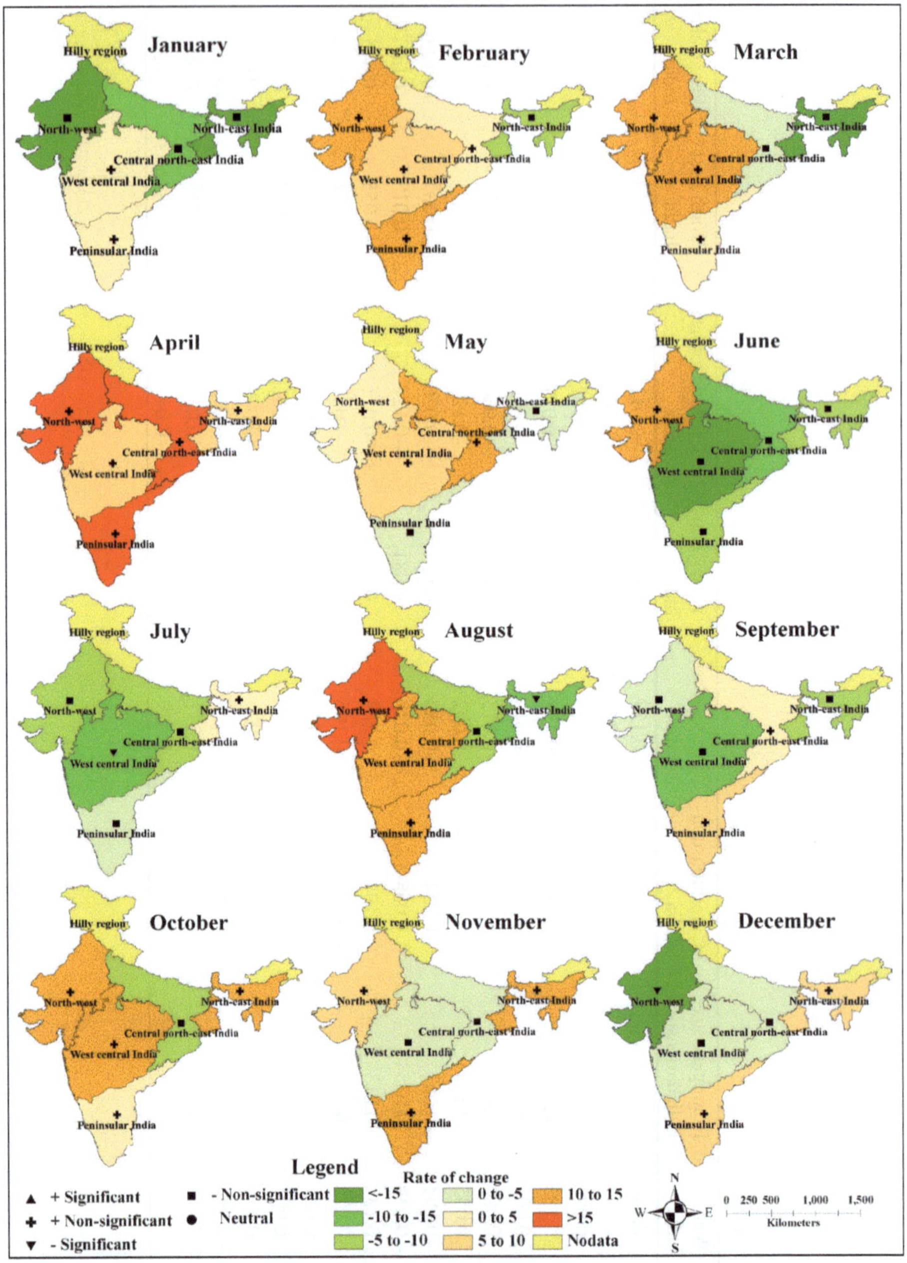

Fig. 4.2 Monthly rainfall trend and rate of change

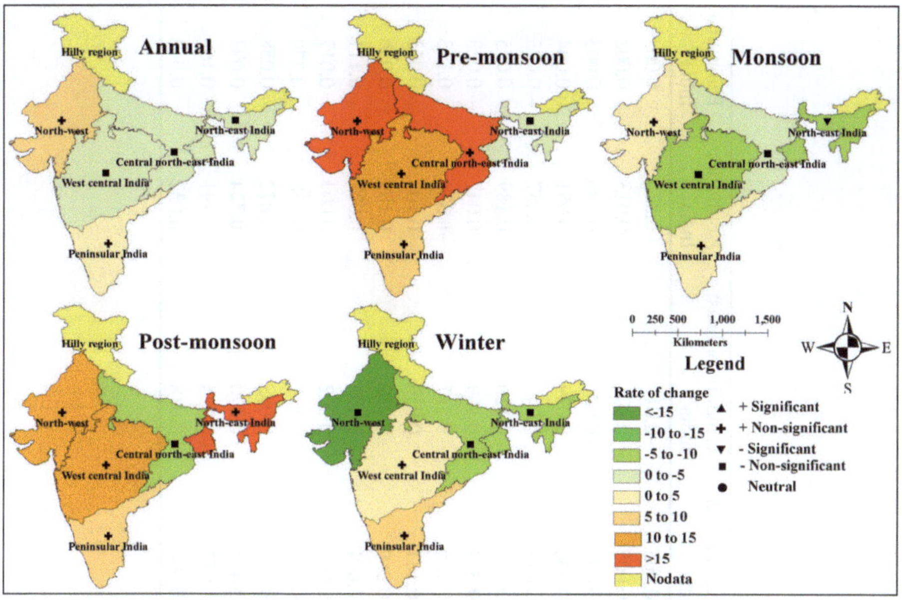

Fig. 4.3 Monthly rainfall trend and rate of change

The seasonal and annual rate of change along with the negative and positive trend values is illustrated in the Fig. 4.3. The monsoon and winter seasons have declining rainfall ranging from −5 % to less than −15 % in 141 years. Annually, highest rate of change in rainfall was observed in North-west India (5–10 %) followed by Peninsular India (0–5 %). West Central, Central North-east and North-east India faced decreasing trend and rate of change (0 to −5 %). Pre-monsoon and post-monsoon periods have increasing rainfall which is evident from the maps, while the maximum reduction was observed in the monsoon and winter seasons particularly in the North-east India.

The rainfall trend magnitude of 141 years is given in Table 4.2. As a whole India reflects a significant positive trend in the month of April (0.036 mm/year) and significant decreasing trend was observed in July (−0.137 mm/year). Rainfall of Homogeneous Indian Monsoon is showing a reduction in rainfall in July (−0.236 mm/year), Core-Monsoon India reflects significant negative rainfall trends in July (−0.287 mm/year), West Central India shows significant reduction in July (−0.314 mm/year) and North East India shows the significant negative rainfall magnitude in August (−0.334 mm/year) and in Monsoon season (−0.724 mm/year). Among all the significant rainfall magnitude results, only the month of April shows the positive trend which is a pre-monsoon month.

The MWP test was done to calculate the change point in these 141 years where the change point for whole India was considered as 1961 (Table 4.3).

The annual total rainfall of 7 regions and whole India is shown in Fig. 4.4 from 1871 to 2011. Linear regression method was used to see the rainfall trend which

Table 4.2 Sen's slope

Station name	Whole India	Homogeneous Indian monsoon	Core-monsoon India	North West India	West Central India	Central Northeast India	North East India	Peninsular India
Jan	−0.011	−0.003	−0.002	−0.010	0.003	−0.012	−0.023	0.003
Feb	0.005	0.006	0.000	0.006	0.005	0.001	−0.013	0.008
Mar	0.003	0.006	0.006	0.004	0.007	−0.001	−0.106	0.003
Apr	**0.036**	0.013	0.004	0.004	0.006	0.039	0.051	0.064
May	0.008	0.011	0.000	0.002	0.011	0.040	−0.057	−0.023
June	−0.102	−0.085	−0.118	0.055	−0.205	−0.120	−0.189	−0.075
July	**−0.137**	**−0.236**	**−0.287**	−0.096	**−0.314**	−0.169	0.047	−0.028
Aug	0.076	0.225	0.258	0.176	0.196	−0.117	**−0.334**	0.123
Sept	−0.078	−0.115	−0.145	−0.017	−0.181	0.030	−0.103	0.063
Oct	0.040	0.048	0.041	0.012	0.060	−0.029	0.139	0.056
Nov	0.025	0.001	0.003	0.002	−0.001	−0.002	0.022	0.111
Dec	−0.004	−0.004	−0.003	−0.009	−0.002	−0.002	0.004	0.017
Annual	−0.043	−0.099	−0.311	0.199	−0.317	−0.165	−0.458	0.379
Pre-monsoon	0.040	0.038	0.006	0.032	0.031	0.088	−0.072	0.058
Monsoon	−0.236	−0.276	−0.409	0.074	−0.435	−0.270	**−0.724**	0.093
Post-monsoon	0.077	0.059	0.051	0.016	0.077	−0.034	0.191	0.168
Winter	0.001	−0.013	−0.023	−0.020	0.000	−0.018	−0.032	0.042

Bold value indicates 95 % significance level

Table 4.3 MWP and SNHT test

	Pettitt's test			SNHT test		
Station	K	t	P value	T_0#	t	P value
Whole India	653	1961	0.89	3.03	1964	0.24
Homogeneous Indian monsoon	688	1964	0.91	3.087	1894	0.25
Core-monsoon India	872	1894	0.98	4.896	1894	0.60
North West India	829	1941	0.97	1.824	1941	0.04
West Central India	1155	1964	1.00	5.006	1964	0.62
Central Northeast India	746	1961	0.94	5.96	2008	0.73
North East India	1195	1956	1.00	13.523	2007	0.99
Peninsular India	723	1914	0.93	6.842	2004	0.82

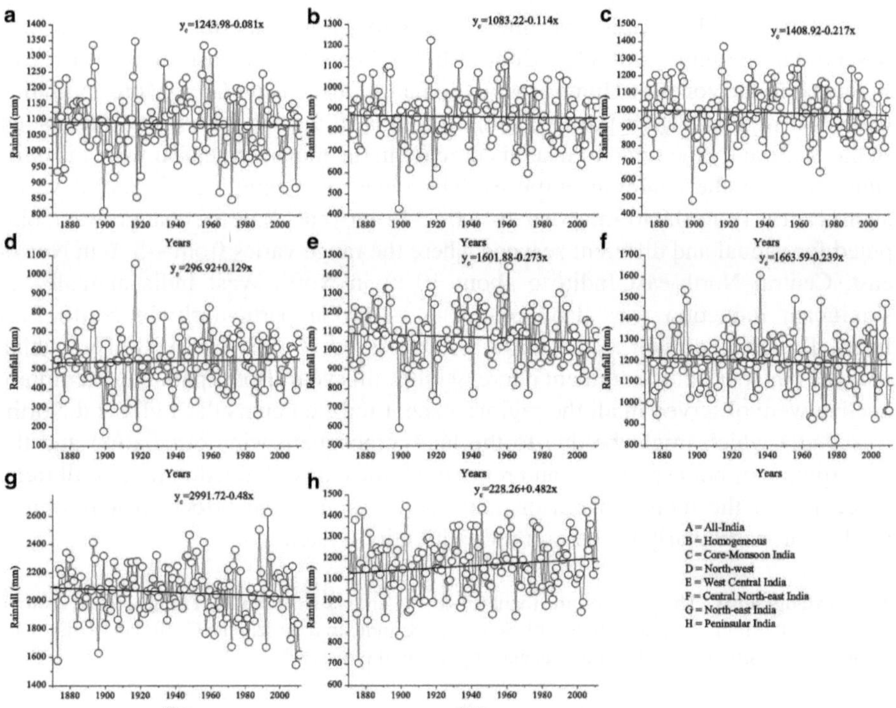

Fig. 4.4 Annual rainfall trend by Linear Regression method from 1871 to 2011

illustrates the greater frequency of decreasing rainfall pattern after 1961 particularly in the North-east, Central North-east and West Central India. However, the increasing rainfall pattern was found in Peninsular India.

The entire rainfall pattern of India indicates significant variability in different regions with some prominent changes in the trend. The overall rainfall trend shows significant decrease in the monsoon months which poses considerable threat to the agricultural economy and will aggravate the demand of water in near future. The change point of 1960–1961 is also significant as the rate of decrease was much more after this period. Decrease in the monsoon of North-east India which was known for its highest rainfall, along with the majority of other regions is also a foremost concern. On the other hand, Peninsular India and North-west India have shown an increasing rainfall trend.

4.5 Conclusion

The study involves the rainfall trend analysis of whole India for 141 years from 1871 to 2011 with MK test and linear regression. Reduction in rainfall trend was observed in the regions of West Central India, Central Northeast India, North-east India and in the zones of Homogeneous Indian Monsoon and Core-Monsoon India. Only two regions depict positive rainfall trends in the Northwest and Peninsular India. Seasonally the trend indicated decrease in the monsoon season which is quite significant for the planning purposes. The range of magnitude of rainfall varies significantly from 0.036 mm/year to −0.334 mm/year. Rate of change was computed for annual and different seasons where the range varies from −5 % in North-east, Central North-east India to about 10 % in North West India annually. A significant reduction was also noticed in monsoon particularly in North East India. The change point in the series was observed in 1961 by MWP Test which indicates the period of dominant decrease after this year. Both spatial and temporal decline were observed in all the regions except for the Peninsular India and North-west India which might be due to the land-atmosphere wind circulation and the contribution of both southwest and retreating monsoons. Reduction in rainfall trend especially in the monsoon season may have a significant effect on agricultural production particularly in the rain fed cultivation system.

Acknowledgments The authors are thankful to the Indian Institute of Tropical Meteorology (IITM) for the rainfall data, Council of Scientific & Industrial Research (CSIR) and University Grant Commission (UGC) for the financial support in the research.

Appendix

Table A1 Mann–Kendall test

Station name	Whole India	Homogeneous Indian monsoon	Core-monsoon India	North West India	West Central India	Central Northeast India	North East India	Peninsular India
Jan	−0.80	−0.19	−0.17	−1.03	0.31	−0.65	−1.30	0.28
Feb	0.35	0.57	0.00	0.79	0.38	0.03	−0.39	0.87
Mar	0.18	0.59	0.97	0.73	0.65	−0.05	−1.48	0.14
Apr	**2.10**	1.29	0.56	0.93	0.47	1.95	0.53	1.67
May	0.25	0.65	0.02	0.09	0.48	0.86	−0.60	−0.33
Jun	−1.22	−0.96	−0.86	0.70	−1.52	−0.83	−1.27	−1.25
Jul	**−2.17**	**−2.00**	**−2.06**	−0.75	**−2.52**	−1.41	0.33	−0.34
Aug	0.85	1.62	1.70	1.20	1.43	−1.09	**−2.72**	1.38
Sep	−0.98	−0.97	−0.94	−0.17	−1.44	0.29	−0.83	0.75
Oct	0.73	0.91	0.67	0.87	0.85	−0.30	1.04	0.43
Nov	0.72	0.12	0.24	1.00	−0.03	−0.10	0.62	0.87
Dec	−0.29	−0.66	−0.63	−1.72	−0.27	−0.37	0.45	0.31
Annual	−0.18	−0.42	−0.95	0.62	−1.15	−0.57	−1.12	1.28
Pre-monsoon	0.98	1.15	0.22	1.38	0.76	1.42	−0.37	0.67
Monsoon	−1.41	−1.07	−1.36	0.25	−1.60	−1.15	**−2.33**	0.50
Post-monsoon	1.10	0.91	0.71	0.79	0.85	−0.32	1.21	0.89
Winter	0.03	−0.52	−0.75	−0.86	0.02	−0.39	−0.58	0.51

Bold value indicates 95 % significance level

Table A2 Rate of change over 141 years

Station name	Whole India	Homogeneous Indian monsoon	Core-monsoon India	North West India	West Central India	Central Northeast India	North East India	Peninsular India
Jan	−14.40	−4.44	−3.16	−20.25	4.89	−11.23	−22.35	3.81
Feb	5.69	9.57	0.05	11.67	7.75	0.42	−6.35	11.60
Mar	2.55	10.55	13.62	10.30	10.77	−1.13	−24.04	2.62
Apr	**19.31**	20.00	8.76	15.89	6.48	33.59	5.44	23.29
May	2.18	9.36	0.32	2.18	7.05	13.32	−3.45	−3.83
Jun	−8.75	−9.34	−10.94	11.88	−17.00	−10.34	−7.09	−6.36
Jul	**−7.13**	**−12.93**	**−13.25**	−7.26	**−14.58**	−7.50	1.67	−2.08
Aug	4.41	14.29	14.03	15.88	10.38	−5.33	**−13.29**	11.03
Sep	−6.47	−11.23	−12.37	−2.95	−13.80	2.00	−5.15	6.02
Oct	7.22	16.60	14.11	13.65	14.21	−5.84	14.04	4.35
Nov	11.14	1.53	2.87	6.31	−0.55	−1.68	11.13	12.77
Dec	−5.34	−9.85	−6.74	−32.35	−3.32	−3.79	6.92	5.82
Annual	−0.56	−1.61	−4.45	5.14	−4.17	−1.94	−3.14	4.60
Pre-monsoon	5.95	15.82	3.21	21.89	10.00	16.79	−2.38	5.90
Monsoon	−3.92	−5.17	−6.54	2.12	−6.64	−3.82	**−7.24**	2.00
Post-monsoon	10.02	15.49	13.06	14.17	13.99	−5.59	16.01	7.76
Winter	0.31	−7.71	−13.90	−15.06	0.16	−6.37	−8.81	9.62

Bold value indicates 95 % significance level

References

Alexandersson H (1986) A homogeneity test applied to precipitation data. J Climatol 6:661–675
Brunetti M, Buffoni L, Maugeri M, Nanni T (2000) Precipitation intensity trends in northern Italy. Int J Climatol 20:1017–1031
Buffoni L, Maugeri M, Nanni T (1999) Precipitation in Italy from 1833 to 1996. Theor Appl Climatol 63:33–40
Burns DA, Klaus J, McHale MR (2007) Recent climate trends and implications for water resources in the Catskill Mountain region, New York, USA. J Hydrol 336:155–170
Cruz RV, Harasawa H, Lal M, Wu S, Anokhin Y, Punsalmaa B, Honda Y, Jafari M, Li C, Huu Ninh N (2007) Asia climate change 2007: impacts, adaptation and vulnerability. In: Parry ML, Canziani OF, Palutikof JP, van der Linden PJ, Hanson CE (eds) Fourth Assessment Report of the Intergovernmental Panel on climate change. Cambridge University Press, Cambridge, pp 469–506
Cunderlik JM, Burn DH (2004) Linkages between regional trends in monthly maximum flows and selected climatic variables. ASCE J Hydrol Eng 9(4):246–256
Englehart PJ, Douglas AV (2006) Defining intraseasonal rainfall variability within the North American monsoon. J Clim 19(17):4243–4253
González JM, Cháidez JJN, Ontiveros VG (2008) Analysis of rainfall trends (1920–2004) in Mexico. Investigaciones Geográficas, Boletín del Instituto de Geografía UNAM 65:38–55
Hamed KH, Rao AR (1998) A modified Mann–Kendall trend test for auto correlated data. J Hydrol 204:182–196
IPCC (Intergovernmental Panel for Climate Change) (2007) Climate Change 2007 – the Scientific Basis, Contribution of Working Group I to the Fourth Assessment Report of the Intergovernmental Panel for Climate Change
Kendall MG (1975) Rank correlation methods. Charles Griffin, London
Khaliq MN, Ouarda TBMJ (2007) Short Communication on the critical values of the standard normal homogeneity test (SNHT). Int J Climatol 27:681–687
Kipkorir EC (2002) Analysis of rainfall climate on the Njemps Flats, Baringo District, Kenya. J Arid Environ 50:445–458
Kruger AC (2006) Observed trends in daily precipitation indices in South Africa: 1910–2004. Int J Climatol 26:2275–2285
Mann HB (1945) Nonparametric tests against trend. Econometrica 13:245–259
Masih I, Uhlenbrook S, Maskey S, Smakhtin V (2010) Streamflow trends and climate linkages in the Zagros Mountains, Iran. Clim Change 104:317–338
Moberg A, Jones PD (2005) Trends in indices for extremes in daily temperature and precipitation in central and Western Europe, 1901–99. Int J Climatol 25:1149–1171
Pettitt AN (1979) A non-parametric approach to the change-point detection. Appl Statist 28(2):126–135
Rodrigo S, Esteban-Parra MJ, Pozo-Vázquez D, Castro-Díez Y (2000) Rainfall variability in southern Spain on decadal to centennial time scales. Int J Climatol 20(7):721–732
Sen PK (1968) Estimates of the regression coefficient based on Kendall's tau. J Am Stat Assoc 63(324):1379–1389
Singh P, Kumar V, Thomas T, Arora M (2008) Changes in rainfall and relative humidity in different river basins in the northwest and central India. Hydrol Processes 22:2982–2992
Storch HV (1993) Misuses of statistical analysis in climate research. Analysis of climate variability: applications of statistical techniques. In: Proceedings of an autumn school organized by the commission of the European Community on Elba from October 30 to November 6, 1993, 2nd ed, Springer, Berlin, 11–26
Suppiah R, Hennessy KJ (1998) Trends in total rainfall, heavy rain events and number of dry days in Australia, 1910–1990. Int J Climatol 10:1141–1164
Theil H (1950) A rank invariant method of linear and polynomial regression analysis, Part 3. Netherlands Akademie van Wettenschappen. Proceedings 53:1397–1412

Xu ZX, Takeuchi K, Ishidaira H (2003) Monotonic trend and step changes in Japanese precipitation. J Hydrol 279:144–150

Yue S, Pilon P, Phinney B (2003) Canadian streamflow trend detection: impact of serial and cross-correlation. Hydrol Sci J 48(1):51–63

Yue S, Hashino M (2003) Long term trends of annual and monthly precipitation in Japan. J Am Water Resour Assoc 39(3):587–596

Chapter 5
Analysing Meteorological and Vegetative Drought in Gujarat

Nairwita Bandyopadhyay and Ashis Kumar Saha

Abstract The impact of drought on vegetation can have significant consequences on livelihood and socio-economic development. Delay in monsoon, high temperature and lack of water resources lead to recurrent droughts in Gujarat. The present work attempts to study the spatio-temporal coverage of drought and its characteristics. Normalized Difference Vegetation Index (NDVI) Anomaly and Rainfall Anomaly Index (RAI) derived through CRU Global Climate dataset and NOAA-AVHRR data respectively for the period 1982–2001 were used for monitoring and comparison of meteorological and vegetative drought situations. Drought patterns, thus delineated, were found to have very good correlation with rainfall. It was observed that both Rainfall Anomaly Index and NDVI Anomaly Index can be used as an indicator for assessment of area affected by meteorological and vegetative drought. The latter showed a high correlation with Rainfall Anomaly Index. The impact of rainfall on vegetation health is, thus, clearly visible. The study was able to delineate the zones more prone to drought with the help of these two indices. This technique proved useful for analysing the spatial and temporal trend of drought, its prevalence, severity level and persistence with the help of freely available meteorological and satellite data. The findings will be of great value for planners and resource managers in quick decision making and forecasting.

Keywords Drought • NDVI Anomaly Index • RAI

N. Bandyopadhyay • A.K. Saha (✉)
Department of Geography, Delhi School of Economics, University of Delhi, Delhi, India
e-mail: ashisksaha@gmail.com

M. Singh et al. (eds.), *Climate Change and Biodiversity: Proceedings of IGU Rohtak Conference, Vol. 1*, Advances in Geographical and Environmental Sciences,
DOI 10.1007/978-4-431-54838-6_5, © Springer Japan 2014

5.1 Introduction

The impact of a natural hazard like drought can be disastrous in a region that has suffered severe droughts multiple times in the past (Bhuiyan et al. 2006). In India, drought has turned into a frequently occurring phenomenon, posing a threat to livelihood and socio-economic development. It also causes significant loss of biodiversity. Increase in water stress affects the agricultural production as well as vegetation health, which leads to extinction of plant and animal species (Kogan 1990). The onset and offset of a drought are difficult to calculate as it accumulates slowly over a period of time unlike other natural hazards like flood, cyclone and earthquake (Nagarajan 2009). Drought can be of various types, meteorological and vegetative being the most frequently occurring ones. Meteorological drought is a function of climatic factors like precipitation, temperature, humidity, etc. Vegetative drought links impacts of meteorological drought to agriculture (Bhuiyan 2004), focusing on precipitation shortages, differences between actual and potential evapo-transpiration, soil water deficits, crop failure, etc.

The most drought-prone subdivisions in western India are Saurashtra-Kutch, Punjab, western Rajasthan, Gujarat plains, Haryana and eastern Rajasthan (Gore and Ray 2002). Gujarat, for example, is an agricultural state where two-thirds of the population is engaged in agriculture and depends heavily on it for livelihood (Gupta and Sharma 2001). To measure the severity of drought, various methods and indices have been developed along with new and advanced GIS techniques (Nagarajan 2009). The satellite remote sensing technique provides baseline data against which future changes can be compared (Chopra 2006), while the GIS technique presents a suitable framework for integrating and analyzing many types of data sources required for disaster monitoring (Burrough 1986).

In the present study, an attempt has been made to analyze temporal and spatial pattern of drought severity in Gujarat using meteorological drought index and vegetative drought index to understand the type of correlation that exists between them. The results obtained with the help of these two indices are presented with maps and statistical methods. The spatio-temporal drought pattern was interpreted through visual observation of the maps generated.

5.2 Study Area

Gujarat, with more than 75 % arid and semi arid area, is exposed to recurrent water scarcity from time to time (ICAR 2005). It is situated (Fig. 5.1) on the western coast of India between 20°06′ N to 24°42′ N latitudes and 68°10′ E to 74°28′ E longitudes. The location is very near to the Thar Desert in Rajasthan and is, thus, very dry and arid in nature. Its topography is characterised by a fertile plain in the south, low hills in the west and mudflats in the north adjoining the Thar Desert.

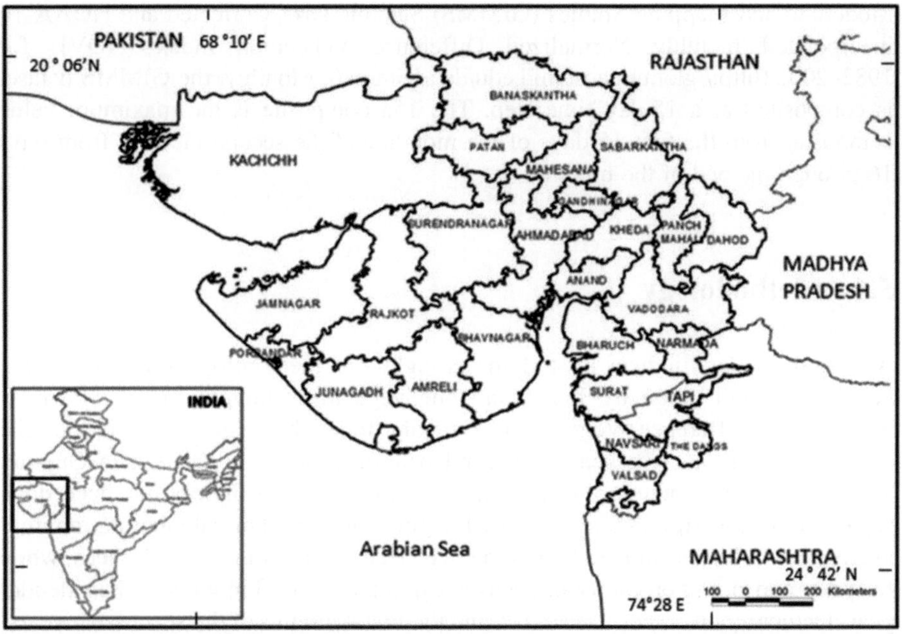

Fig. 5.1 Location map of Gujarat, India

The climate of Gujarat is unique as it ranges from dry desert areas to high altitude rain forests. According to variability of rainfall, actual evapo-transpiration and other factors, Gujarat can be divided into seven major eco-regions by National Bureau of Soil Survey and Land Use Planning (ICAR 2005). These are characterised by erratic and uneven rainfall in some areas while heavy rain in others. Gujarat receives on an average only one rainfall a year from the south west monsoon, which ranges from 1,000–2,000 mm in the southern rocky highland to 250–400 mm in the Kutch area. The surface water is concentrated mainly in the southern and central parts of the state. The two sub regions of Gujarat are Saurashtra and Kutch. While Saurashtra is characterised by semi arid, hard rock regions with an annual rainfall of 400 mm, Kutch has an annual rainfall of around 300 mm and is the most arid part of Gujarat (Gore and Ray 2002). The state faces severe water crisis mostly in the northern parts. It has more than 80 % of agriculture dependent on groundwater.

5.3 Data Source

The CRU TS 2.1 Global Climate dataset with resolution of 0.5° (approximate area 90 km), containing precipitation data in mm is taken for the calculation of Rainfall Anomaly Index. This dataset is available for free download at: http://csi.cgiar.org/cru/. The data set for vegetative drought calculation is obtained from Global Inventory

Modelling and Mapping Studies (GIMMS) Satellite Drift Corrected and NOAA-16 incorporated monthly Normalized Difference Vegetation Index (NDVI) for 1982–2002 (http://glcf.umiacs.umd.edu/data/gimms/). Monthly, the GIMMS dataset is composited at a 15-day time step. The 15a composite is the maximum value composite from the first 15 days of the month, and the second (15b) is from days 16 through the end of the month.

5.4 Methodology

Among the several indices in studying drought, one meteorological drought index namely Rainfall Anomaly Index and one vegetative drought index known as Normalized Difference Vegetation Anomaly Index have been selected in the present study to analyse spatio-temporal pattern of seasonal drought in Gujarat. The calculation of meteorological and vegetative drought indices has been done separately using ArcGIS and ERDAS Imagine software. For calculation, monthly average rainfall from June to September has been taken for monsoon indices, while for post-monsoon (non-monsoon) season calculation, last 3 months of the calendar year- October, November, December and the first 4 months of the successive years- January, February, March, April and May are taken into account for 20 years (1982–2001). A temporal as well as spatial trend analysis of drought in Gujarat was also undertaken for this duration, taken as a common period for the two drought indices calculation and analysis. To understand the relationship of these two indices with rainfall, Product Moment Correlation Function was used with the help of SPSS software. A brief detail of the calculation of the two drought indices selected is as follows:

5.4.1 Rainfall Anomaly Index

This index was devised by Van Rooy (1965), and was further used by Chopra (2006) and [Bandyopadhyay (2012) Drought severity analysis using geospatial data in Gujarat. M.Phil. Dissertation, University of Delhi, Unpublished] for identifying meteorological drought in places where long-term average rainfall is less followed by high variability in rainfall, increasing the likelihood of drought. The utility of using this index is that it can be computed for different time scales and can be used for future prediction of drought occurrences. The accumulated deficit (RAI < 0 %) or surplus (RAI > 0 %) of precipitation over a specified period, and a normalized measure of relative precipitation anomalies at multiple time scales can be captured with its use.

Rainfall anomaly is computed by dividing the difference between the seasonal rainfall for 'ith' year from mean seasonal rainfall divided by the mean seasonal rainfall. Rainfall anomaly has been computed from 1982–2001 for monsoon and non-monsoon seasons to indicate meteorological drought. Rainfall anomaly is computed as:

$$\mathrm{RFAi} = [(\mathrm{RFi} - \mathrm{RF}\mu)/(\mathrm{RF}\mu)] * 100 \tag{5.1}$$

where, RFAi is rainfall anomaly for ith year; RFi is seasonal rainfall for ith year and RFμ is mean seasonal rainfall. The values range from −100 to +100 % departure from normal. The value above 0 % is a situation when there is no drought, negative departure of 0–10 % denotes mild drought, 10–20 % is moderate drought, 20–30 % is severe drought and negative departure of 30–100 % denotes extreme drought condition (Chopra 2006). The area under each category is then calculated by dividing the number of pixels falling under a particular category by the total number of pixels, and then multiplying it by 100.

5.4.2 NDVI Anomaly Index

The severity of a drought (or the extent of wetness on the other end of the spectrum) may be defined as NDVI deviation from its long-term mean NDVI Anomaly (Anyamba and Tucker 2005). The formula for calculation is as follows:

$$\mathrm{NDVI\ Anomaly} = \mathrm{NDVIi} - \mathrm{NDVI\ mean, m} \tag{5.2}$$

Where NDVIi is the NDVI value for month i and NDVI mean, m is the long-term mean NDVI for the same month m (e.g., in a data record from 1982 to 2001, 20 long- term NDVI means—one for each monsoon/non-monsoon month of a particular year). When NDVI Anomaly is negative, it indicates the below-normal vegetation condition/health and, therefore, suggests a prevailing drought situation (Goetz 1997). High negative departure indicates greater stress on vegetation. In general, the departure from the long-term mean NDVI is effectively more than just a drought indicator, as it would reflect the conditions of healthy vegetation in normal and wet months/years. Its limitations are that the deviation from the mean does not take into account the standard deviation, and hence can be misinterpreted when the variability in vegetation conditions in a region is very high in any one given year. The values range from −100 to +100 % departure from normal. Vegetative drought is a result of gradual accumulation of vegetation stress. The NDVI reflects the nature of vegetation through the ratio of response in near Infrared (Ch 2) and visible (Ch 1) bands of NOAA-AVHRR.

$$\mathrm{NDVI} = (\mathrm{Ch2} - \mathrm{Ch1})/(\mathrm{Ch2} + \mathrm{Ch1}) \tag{5.3}$$

Mean NDVI for 20 years was then computed by using the following expression:

$$\mathrm{Mean\ NDVI} = (\mathrm{NDVI\ max\ y1} + \mathrm{NDVI\ max\ y2} + \ldots\ldots + \mathrm{NDVI\ max\ 20})/20 \tag{5.4}$$

Where, NDVI max stands for the representative NDVI value for monsoon/non-monsoon months of a particular year.

After taking NDVI max from1981–2000, an average of these NDVI max images was computed to get the mean NDVI max values during past 20 years. Anomaly was then computed as:

$$\text{Anomaly NDVIi} = (\text{NDVI max i} - \text{mean NDVI max})/(\text{mean NDVI max}) * 10 \tag{5.5}$$

where, Anomaly NDVIi is NDVI anomaly in ith year, NDVI max is maximum representative NDVI and mean NDVI max is the average of maximum NDVI during 1982–2002.

The values from 0 % to −10 % equivalent NDVI anomalies have been classed into mild drought, −10 % to −20 % equivalent NDVI anomalies as moderate drought, −20 % to −30 % equivalent NDVI anomalies as severe drought and −30 % to −100 % equivalent NDVI anomalies as extreme drought.

5.5 Drought in Gujarat

5.5.1 Meteorological Drought

Meteorological drought is usually based on long-term precipitation departures from normal, but there is no consensus regarding the threshold of the deficit or the minimum duration of the lack of precipitation that makes a dry spell an official drought. This happens when the actual rainfall in an area is significantly less than the climatological mean of that area. The impact of rainfall deficiency on drought development in Gujarat has been quantified with the help of calculation of the precipitation deficit in the monsoon and the non-monsoon periods since 1982–2002, with the help of Rainfall Anomaly Index. Rainfall anomaly maps indicate that meteorological drought in Gujarat appears randomly in both monsoon and non-monsoon seasons. Inconsistency in rainfall distribution has led to irregular pattern of drought in this region. The picture of meteorological drought was observed to change consistently in monsoon and non-monsoon as it was dependent on rainfall amount and its distribution spatially. It was observed that the region experienced a continuous spell of dry conditions from 1984 to 1985 non-monsoon till 1987 monsoon (Fig. 5.2) after which the region became drought-free. In approximate terms, it can be identified from visual interpretation that the north western and south western sectors are more drought-prone compared to the eastern and south eastern pockets. The spatial pattern of rainfall anomaly also identified some pattern of drought-prone region but delineation of drought zones was not possible as it was not persistent to any particular zone or sector for more than two consecutive monsoon or non-monsoon seasons.

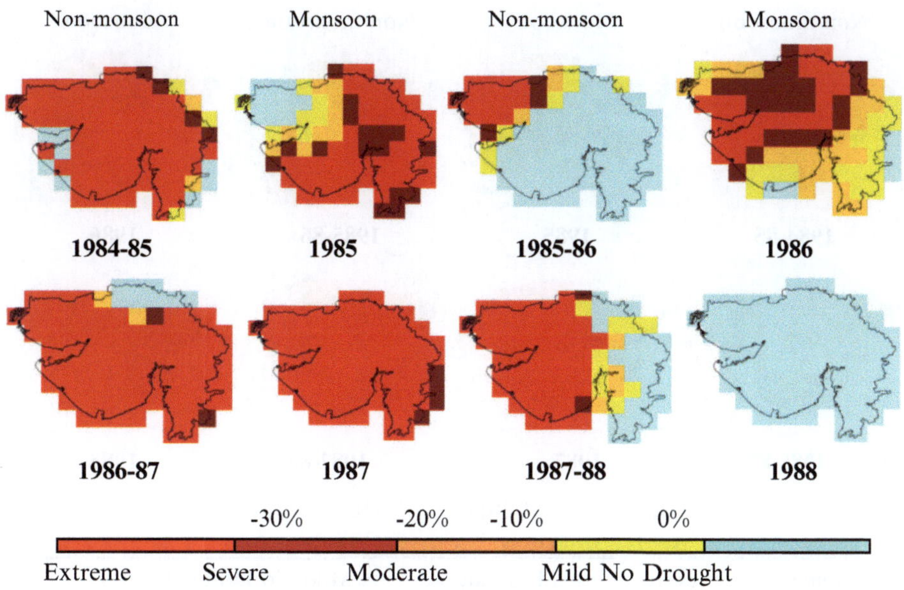

Fig. 5.2 Rainfall anomaly drought dynamics from 1984 to 1988

5.5.2 *Vegetative Drought*

Vegetative Drought Assessment considers all vegetation: natural vegetation as well as agricultural crops. It occurs when there is insufficient soil moisture to meet the need of a particular crop at a particular point in time. Deficit rainfall over cropped areas during their growth cycle can lead to destruction of crops. Extremely dry climate, drought vegetation and agriculture that is possible only by irrigation, are characteristic features of vegetative drought. During this type of drought, soil moisture and rainfall are inadequate during the crop growing season to support its healthy growth to maturity, which causes extreme crop stress and wilting. When drought begins, the vegetative sector is the first to be affected because of its heavy dependence on soil water which depletes rapidly during extended dry periods.

In order to study vegetative drought in this paper, NDVI anomalies have been computed to define the severity of vegetative drought for both monsoon and non-monsoon months. From 1982 to 2001, 20 maps were generated and classified into these different classes.

The monsoon and non-monsoon season of 1982 and 1983 showed considerable deviation of NDVI values. During 1984 and before 1985, Gujarat shows normal vegetation with NDVI values showing no negative departure. During the monsoon of 1985, it can be observed that vegetation underwent stress and health of vegetation declined as negative departure of $>10\ \%$ was observed. It experienced a continuous spell of drought till the monsoon of 1988 owing to poor rainfall in three consecutive monsoons and one intermediate monsoon season. Gujarat encountered the worst

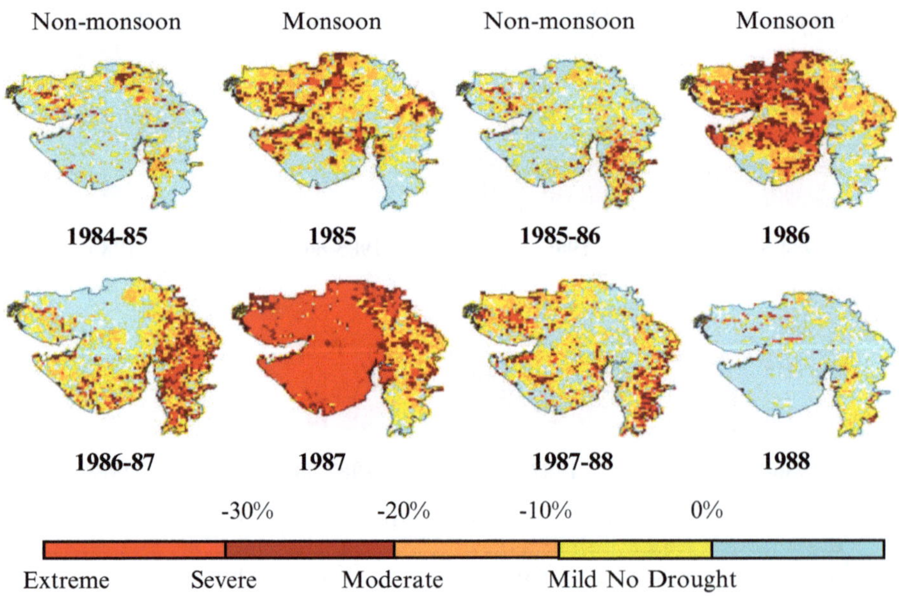

Fig. 5.3 NDVI anomaly drought dynamics from 1984 to 1988

situation in 1987 monsoon when the whole region suffered severe to extreme NDVI anomaly (−30 to −100 %) due to heavy scarcity of rainfall. The region remained almost drought-free after 1987 (Fig. 5.3) in the following years and mild to moderate drought appeared in certain seasons. During 1991, drought appeared in the north western pocket and in 1993, the magnitude increased with NDVI values showing more than 30 % negative departure. During 2000 monsoon, the southern part was again affected due to deficient monsoon as only 14 % of the area received normal rain (Gupta and Sharma 2001). The loss to Kharif crops is estimated to be 36 lakh metric tons which is about 43 % of the production level of 1998–1999 (Gupta and Sharma 2001).

Comparison with the agricultural landuse map (ICAR 2005) reveals that the areas that are affected by drought in monsoon are the areas of rain-fed crops (northern, central and eastern parts), and the pockets that are affected during non-monsoon months are areas of irrigated crops (north eastern, southern and south western parts). Vegetative Drought is, thus, crop-specific and varies according to the crops' ability to cope with water stress.

5.6 Comparison and Correlation

The chief parameter for meteorological drought is precipitation while vegetative or vegetative drought is a manifestation of meteorological drought. The drought maps that are generated from 1982–2001 are correlated to one another. During both monsoon and non-monsoon periods of 1986–1987, lack of sufficient rainfall

Table 5.1 Product moment correlation coefficient of RAI and NDVI anomaly

Index	Monsoon	Non-monsoon
Rainfall Anomaly Index	−0.923	0.140
NDVI Anomaly Index	−0.715	−0.048

Table 5.2 Area affected due to drought during monsoon and non-monsoon

Index	Monsoon (Area in %)	Non-monsoon (Area in %)
Rainfall Anomaly Index	54	61
NDVI Anomaly Index	48	50

resulted in extreme meteorological and vegetative drought in almost the entire state of Gujarat, barring some pockets. Meteorological drought development resulted in acute vegetative drought during the monsoon of 1987. During 1990–1991, poor rainfall during non-monsoon and monsoon seasons lead to stress of vegetation in the north western sector. Water deficit during non-monsoon in the western part, north east and central pockets led to vegetation stress in the monsoon of 2001 in north western, eastern and central pockets. It is, thus, the cumulative effect of water deficit during both non-monsoon and monsoon season that leads to vegetation stress during the coming monsoon season and not due to water deficit only during the monsoon months.

5.6.1 Relationship with Rainfall

A negative correlation between rainfall and the area affected by drought can be observed with the help of Pearson's Product Moment correlation (Kumar and Panu 1997) coefficients from 1982–2002. During the monsoon season, meteorological drought showed negative correlation but for vegetative drought the correlation is true for both monsoon and non-monsoon seasons. No significant correlation was seen in case of meteorological drought during non-monsoon season as rainfall was very minimum, and water deficit too was not remarkable.

During monsoon, with increase in rainfall, decrease in drought affected area was seen. NDVI anomaly drought shows that the correlation coefficient value is −0.715 (Table 5.1) during monsoon and −0.048 during non-monsoon. This indicates vegetation health also decreases with decrease in rainfall, which leads to increase in area affected by poor vegetation under water stress. It can be observed that the area affected under meteorological drought is more compared to the vegetative drought during monsoon as well as non-monsoon season (Table 5.2).

Further, it was seen that when rainfall was very low, area affected by drought identified by the two indices was high and with higher rainfall, the area affected identified by the two indices was low (Fig. 5.4). The increase in area under vegetative drought follows the increase in area affected by meteorological drought. Thus, it can be inferred that vegetative drought follows a meteorological drought.

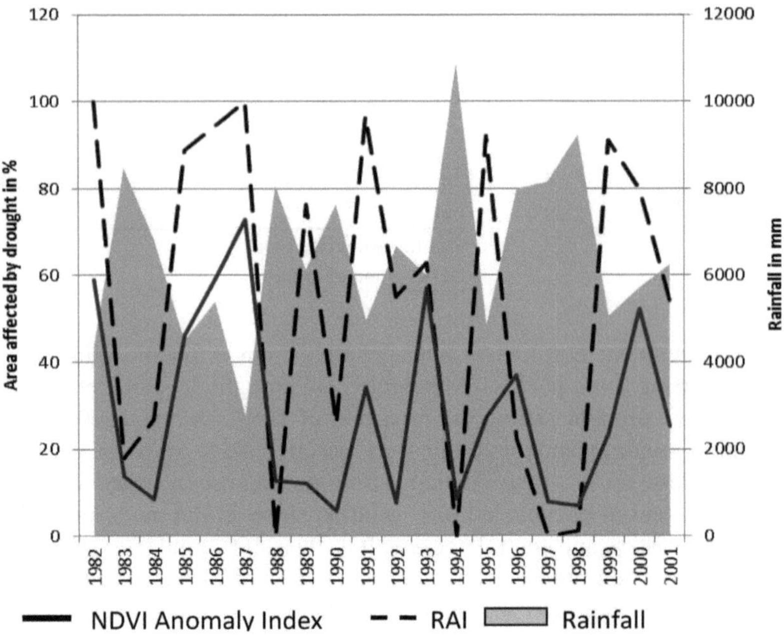

Fig. 5.4 Relationship of RAI and NDVI anomaly with rainfall (monsoon 1982–2001)

5.7 Conclusion

Rainfall records initially show deviation from normal rainfall pointing towards a drought situation. After that, soil moisture begins to decrease which in turn affects vegetation. The drought maps of Rainfall Anomaly Index and NDVI Anomaly Index show that both meteorological and vegetative drought occur in Gujarat frequently, but in an irregular manner. Rainfall being unevenly distributed, drought zonation of Gujarat based on only meteorological parameters, is not possible. Monsoon and non-monsoon seasons both have experienced meteorological drought. Though maximum of the rainfall occurs in monsoon, non-monsoon season also showed deviation of rainfall than normal in certain years.

Vegetation pattern also does not show any spatio-temporal trend. The vegetative health maps indicate that vegetation growth can withstand adverse meteorological conditions for several seasons to maintain good vegetation health, in spite of unfavourable meteorological conditions. It can be observed that vegetative drought is slow to begin but quicker to withdraw.

It is also observed that there exists a negative correlation between rainfall and the two drought indices. The area affected under meteorological drought is more compared to vegetative drought as calculated by the indices during monsoon and non-monsoon season (Table 5.2). Due to decrease in rainfall, the rise in area affected during vegetative drought follows meteorological drought, showing a drag effect.

Meteorological drought does not necessarily affect the actual vegetation health and thus, only on the basis of meteorological parameters, it is incorrect to define a drought situation. Vegetation health maps or calculation of vegetative drought index determines the impact of water deficit on vegetation that generally leads to crop failure, poor health of vegetation and eventually drought. Therefore, Rainfall Anomaly Index and NDVI Anomaly Index together present a better picture of drought, aiding in its early forecast, especially in Gujarat where rainfall is uncertain and unevenly distributed, abetted by a dry and irregular terrain.

Acknowledgments NB is thankful to University Grant Commission (UGC), India for awarding Junior Research Fellowship to carry out this research.

References

Anyamba A, Tucker CJ (2005) Analysis of Sahelian vegetation dynamics using NOAA-AVHRR NDVI data from 1981–2003. Arid Environ 63(3):596–614

Burrough PA (1986) Principles of geographic information systems for land resources assessment. Oxford, New York

Bhuiyan C (2004) Various drought indices for monitoring drought condition in Aravalli terrain of India. In: Proceedings of the XXth ISPRS conference. Int Soc Photogramm Remote Sens, Istanbul. http://www.isprs.org/istanbul2004/comm7/papers/243.pdf. Accessed 4 March 2011

Bhuiyan C, Singh RP, Kogan FN (2006) Monitoring drought dynamics in the Aravalli region (India) using different indices based on ground and remote sensing data. Int J Earth Obs Geoinf 8:289–302

Chopra P (2006) Drought risk assessment using remote sensing and GIS: a case study of Gujarat. http://www.mesa7a.com/up/mscphd/MSC%20GIS_RS%20Risk%20Ass-esment%202006.pdf). Accessed 3rd April 2011

Goetz SJ (1997) Multi-sensor analysis of NDVI, surface temperature and biophysical variables at a mixed grassland site. Int J Remote Sens 18:71–94

Gore PG, Ray KCS (2002) Drought aridity over districts of Gujarat. J Agromet 75–85

Gupta LC, Sharma K (2001) Drought in Gujarat 1999–2000 &2000–2001. Indian Institute of Public Administration, New Delhi, pp 9–47

ICAR (2005) Handbook of agriculture. National Bureau of Soil Survey and Land Use Planning, Indian Council of Agricultural Research, New Delhi, p 1302

Kogan FN (1990) Remote sensing of weather impacts on vegetation in non-homogeneous areas. Int J Remote Sens 11(8):1405–1419

Kumar V, Panu U (1997) Predictive assessment of severity of agricultural droughts based on agro-climatic factors. J Am Water Resour Assoc 33(6):1255–1264

Nagarajan R (2009) Drought assessment. Capital Publishing Company, New Delhi

Van Rooy MP (1965) A Rainfall anomaly index (RAI) independent of time and space. Notos 14:43–48

Chapter 6
Multiple Linear Regression Based Statistical Downscaling of Daily Precipitation in a Canal Command

Prabhash Kumar Mishra, Deepak Khare, Arun Mondal, and Sananda Kundu

Abstract The climate impact studies, particularly in hydrology, often require climate information at fine scale for present as well as future scenario. Global Climate Model (GCM) estimates climate change scenarios on coarse spatial resolution. Therefore, different techniques have been evolved to downscale the coarse-grid scale GCM data to finer scale surface variables of interest. In the present study, the Statistical Downscaling Model (SDSM) has been applied to downscale daily precipitation from simulated GCM data. SDSM utilizes Multiple Linear Regression (MLR) technique. The daily precipitation data (1961–2001) representing Tawa region has been considered as input (predictand) to the model. The model has been calibrated (1961–1991) and validated (1992–2001) with screened large-scale predictors of (National Centre for Environmental Prediction (NCEP) reanalysis data. The prediction of future daily rainfall for the study area has been carried out for the period 2020s, 2050s and 2080s corresponding to HadCM3 A2 variables. The calibration and validation results confirm the SDSM model acceptability slightly at a lower degree. The results of the downscaled daily precipitation for the future period indicate an increasing trend in the mean daily precipitation.

Keywords GCM data • Scenario generation • SDSM • Statistical downscaling • Tawa command

P.K. Mishra (✉)
Department of Water Resources Development and Management,
Indian Institute of Technology Roorkee, Roorkee, India

Scientist, Water Resources Systems Division, National Institute of Hydrology,
Roorkee, India
e-mail: erprabhash@gmail.com

D. Khare • A. Mondal • S. Kundu
Department of Water Resources Development and Management,
Indian Institute of Technology Roorkee, Roorkee, India
e-mail: kharefwt@gmail.com; arun.iirs@gmail.com; sanandakundu@gmail.com

M. Singh et al. (eds.), *Climate Change and Biodiversity: Proceedings of IGU Rohtak Conference, Vol. 1*, Advances in Geographical and Environmental Sciences,
DOI 10.1007/978-4-431-54838-6_6, © Springer Japan 2014

6.1 Introduction

Climate is a complex system, and is very difficult to quantify its' variables. Precipitation is an important parameter (variable) for climate change impact studies. A proper assessment of precipitation for past events, and its' future scenarios is needed for water resources planning. Global Circulation Models (GCMs) are tools available to simulate the ongoing and future changes in climate at global scale. GCMs are numerical models representing the physical processes of the earth-atmosphere-ocean system (Robock et al. 1993; Hewitson and Crane 1996; Wilby and Wigley 1997; Prudhomme et al. 2003; Crawford et al. 2007). These models are of coarse-grid resolution, and of high accuracy at large spatial scales (Bardossy 1997; Ojha et al. 2010; Hassan and Harun 2012). However, impact studies by hydrologists and water resources planner require local/regional-scale hydrological variability to represent local climate phenomena. Hence, different approaches are evolved to downscale the coarse-grid scale GCM data to finer scale surface variables in last few decades. Such methods include canonical correlation analysis, multiple linear regressions, artificial neural networks and support vector machines (Murphy 2000; Lall et al. 2001; Huth 2002; Aksornsingchai and Srinilta 2011; Ghosh and Mujumdar 2006; Raje and Mujumdar 2009; Ghosh 2010; Kannan and Ghosh 2010; Raje and Mujumdar 2011; Hashmi and Shamseldin 2011; Kodra et al. 2012). Recently, downscaling of precipitation has found wide utility for scenario generation on different time scales. SDSM is one of the statistical downscaling tools that implement the multiple linear regression model, and provides scenario of daily surface weather variables under the present and future climate forcing. The tool also performs ancillary tasks of data quality control and transformation, prescreening of predictor variables, model calibration and validation, scenario generation, statistical analysis and its representation of climate data (Wilby and Dawson 2007).

The objective of this study is to understand and utilize Linear Multiple Regression (MLR) technique to downscale mean daily precipitation, both for present and future, for crop planning over a command area corresponding to HadCM3 A2 GCM data utilizing SDSM model. The running paper is subsequently structured as follows: Sect. 6.2 provides a brief description of the study area, followed by discussion on data used in Sects. 6.3 and 6.4 describes the procedure to screen predictor variables for downscaling, and the proposed methodology for development of the regression based model for downscaling precipitation for the command area. Section 6.5 presents the results with discussion, and finally, Sect. 6.6 provides the conclusions drawn from the study.

6.2 Study Area

Tawa command is spread over in an area of about 5,273.12 km^2 falling in the district of Hoshangabad, Madhya Pradesh, India. It lies between 22°54′ N to 23°00′ N latitude and 76°457′ E to 78°45′ E. The area is characterized by a hot summer and

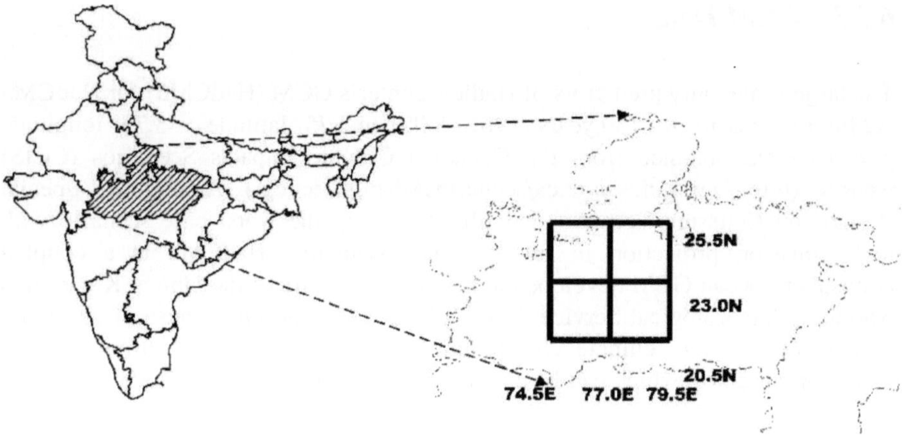

Fig. 6.1 Location map of the Tawa command with NCEP grid (2.5° × 2.5°)

evenly distributed rainfall during the southwest monsoon period. The temperature starts rising from beginning of February and peak is reached in the month of May touching the mercury at 42 °C (Normal). The winter season commences with November and temperature dips to 7.2 °C in the month of December. The relative humidity during summer is low in the month of April i.e. about 18.1 % and is maximum in August i.e., 86.7 %. The annual rainfall varies from 652 mm to 1,898 mm in the command area with average of 1,175 mm based on observations recorded during 1961–2010. The location map of the study area is shown in Fig. 6.1.

6.3 Data Used

6.3.1 Meteorological Data

The daily precipitation data were collected from India Meteorological Department (IMD), Pune for the periods 1961–2001. The daily data were converted to monthly, seasonal and annual time scale before analysis is done.

6.3.2 Reanalysis Data

The daily observed predictor data (re-analysis data) of atmospheric variables, derived from the National Center of Environmental Prediction (NCEP) on 2.5° latitude × 2.5° longitude grid-scale for 41 years (1961–2001) are obtained from the Canadian Climate Impacts Scenarios (CCIS) website (http://www.cics.uvic.ca/scenarios/sdsm/select.cgi).

6.3.3 GCM Data

The large-scale daily predictors of Hadley Center's GCM (HadCM3) for HadCM3 A2 future scenarios for 139 years (1961–2099) on 3.75° latitude × 3.75° longitude grid-scale are obtained from the Canadian Climate Impacts Scenarios (CCIS) website (http://www.cics.uvic.ca/scenarios/sdsm/select.cgi). Among the Special Reports on Emission Scenarios (SRES) A2, being the worst case scenario with high emission projection in future, was considered. HadCM3 is a coupled atmosphere-ocean GCM developed at the Hadley Centre of the United Kingdom's National Meteorological Service. HadCM3 has been chosen because of its' wider acceptance in many climate change impact studies. Further, it provides daily predictor variables, which can be exclusively used for the SDSM model.

6.4 Methodology

The Statistical Downscaling Model (SDSM) is a multiple regression-based tool for generating future scenarios to assess the impact of climate change. It has the ability to capture the inter-annual variability better than other statistical downscaling approaches, e.g. weather generators, weather typing. The model requires two types of daily data, the first type corresponds to local data known as 'Predictand' (Precipitation, temperature) and the second type corresponds to large-scale data of different atmospheric variables known as 'Predictors'(NCEP reanalysis data and simulated GCM based data), for downscaling. Formulating an empirical relationship between predictand and predictor is central to the downscaling technique. This can be achieved by methods, both parametric (Multiple Linear Regression) and non-parametric (Artificial Neural Network; Support Vector Machine). The study has been carried out using SDSM tool version 4.2.9.

6.4.1 Selection of Predictors

For downscaling predictand, the selection of suitable predictors is one of the most important and time consuming steps during downscaling. The appropriate predictor variables are selected through scatter plots, correlation and partial correlation analysis performed between the predictand of interest and predictors. The observed daily NCEP reanalysis data set for the periods 1961–2001 was used to identify the predictors.

6.4.2 Model Calibration and Validation

Model calibration involves development of an empirical relationship, here multiple linear regression, between the predict and of interest and identified daily observed predictors. Part of the NCEP reanalysis data for the period 1961–1991 is used for model calibration, and remaining data between 1992 and 2001 for validation. Validation process enables to produce synthetic daily data based on inputs of the data not considered during model calibration and the formulated regression model. The model performance was evaluated based on the coefficient of correlation (R), defined as:

$$R = \frac{\sum \left(X_{obs} - \overline{X}_{obs}\right)\left(X_{\text{mod}} - \overline{X}_{\text{mod}}\right)}{\sqrt{\sum \left(X_{obs} - \overline{X}_{obs}\right)^2 \sum \left(X_{obs} - \overline{X}_{obs}\right)^2}} \tag{6.1}$$

where,

X_{obs} = Observed value; $\overline{X}_{obs}$ = Mean observed value ; X_{mod} = Modelled value; $\overline{X}_{\text{mod}}$ = Mean modelled value.

6.4.3 Scenario Generation

The validated regression model is applied to generate future scenario for the region utilizing the simulated HadCM3 A2 GCMs data. The study assumes that the relationship between predictor and predictand remains valid under the future climate conditions. Twenty ensembles of daily synthetic precipitation for a period of 139 years (1961–2099) have been generated. The ensemble values are averaged and divided into three separate time period viz. 2020s (2011–2040), 2050s (2041–2070) and 2080s (2071–2099).

6.5 Results and Discussions

6.5.1 Selection of Predictor Variables

The selection of predictor variables is the most significant and time consuming step in statistical downscaling. A list of predictor variables (NCEP and GCM) of a grid-box closest to the Tawa region is presented in Table 6.1. A total of 26 large-scale predictor variables have been considered in the initial screening process. These are categorized into six types based on the atmospheric pressure level. The predictors are selected based on correlation and partial correlation analysis of NCEP predictors and observed weather variables for the period 1961–2003 in SDSM. Variables with higher correlation coefficients between precitand (precipitation) and predictors

Table 6.1 Name and description of all NCEP and GCM predictors

Sl No.	Atmospheric pressure level	NCEP variables	Name	Unit
A.	1013.25 hPa (1)	MSL pressure	mslp	Pa
B.	1000 hPa (6)	Wind speed (Geostrophic)	p_f	m/s
		Zonal (Eastward) velocity (U-component)	p_u	m/s
		Meridional (Northward) velocity (V-component)	p_v	m/s
		Vorticity	p_z	s^{-1}
		Wind direction	p_th	degree
		Divergence	p_zh	s^{-1}
C	850 hPa (8)	Wind speed (Geostrophic)	p8_f	m/s
		Zonal (Eastward) velocity (U-component)	p8_u	m/s
		Meridional (Northward) velocity (V-component)	p8_v	m/s
		Vorticity	p8_z	s^{-1}
		Wind direction	p8_th	degree
		Divergence	p8_zh	s^{-1}
		Geopotential height	p850	m
		Relative humidity	r850	%
D	500 hPa (8)	Wind speed (Geostrophic)	p5_f	m/s
		Zonal (Eastward) velocity (U-component)	p5_u	m/s
		Meridional (Northward) velocity (V-component)	p5_v	m/s
		Vorticity	p5_z	s^{-1}
		Wind direction	p5_th	
		Divergence	p5_zh	s^{-1}
		Geopotential height	p500	m
		Relative humidity	r500	%
E	Near surface (3)	Specific humidity	shum	g/kg
		Mean temperature	temp	°C
		Relative humidity	rhum	%

(NCEP) are chosen for model formulation for scenario generation. The selected predictors with their corresponding correlation coefficients, partial correlation and p value are given in Table 6.2. The scatter plot between selected predictors and observed variable are shown in Fig. 6.2. These statistics help to identify the amount of explanatory power that is unique to each predictor. A 5 % significance level ($p < 0.05$) is used to test the significance of predictor-predictand correlation.

6.5.2 Model Calibration and Validation Results

The model calibration process formulates downscaling model based on multiple regressions between the predictand (observed precipitation) and selected NCEP predictors (Table 6.2). Since the predictand-predictor relationship is governed by wet-day occurrence, an intermediate process in the case of precipitation, a threshold value of 0.3 mm rainfall is considered during model calibration. Calibration (1961–1991)

Table 6.2 Selected NCEP predictors with correlation coefficient, partial correlation and p value

Sl No.	Selected predictors	Correlation coefficients	Partial correlation	P value
1	ncepp_zas	0.374	0.128	0.0001
2	ncepp5_zas	0.320	0.089	0.0001
3	ncepp8_zas	0.344	0.062	0.0006
4	ncep_mslp_as	−0.214	−0.058	0.0013
5	ncepp850as	−0.239	0.056	0.0022
6	ncep_shum_as	0.174	−0.047	0.0105
7	ncep_rhum_as	0.172	0.041	0.0285

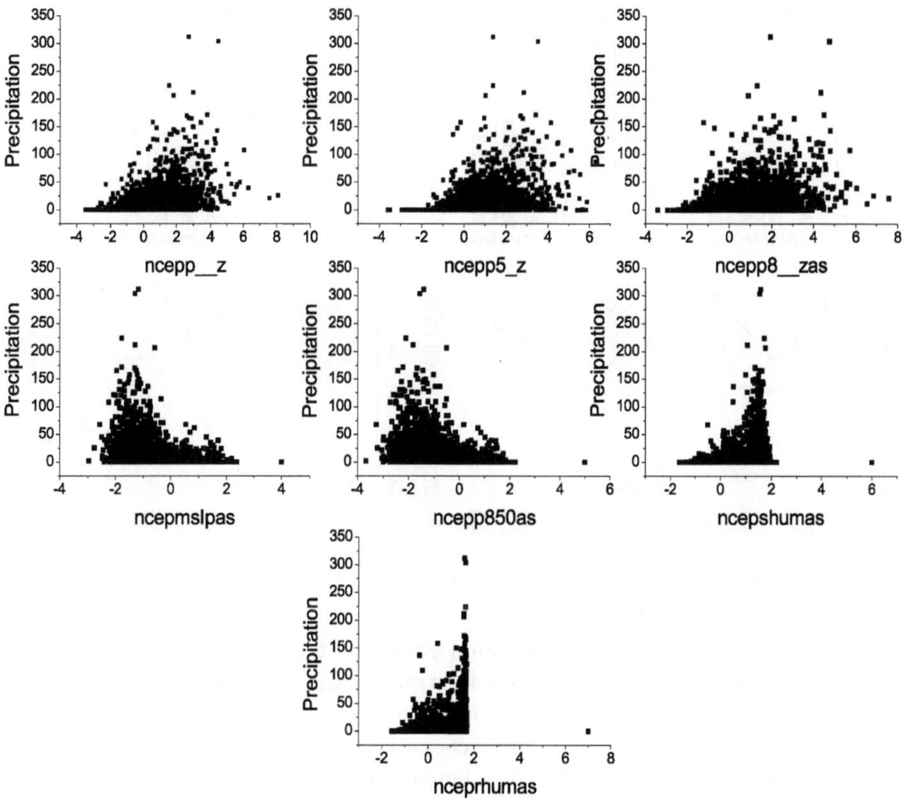

Fig. 6.2 Scatter plots between predictand and selected NCEP predictors

and validation (1992–2001) result of the model downscaling (1961–1991) of daily rainfall is presented in Table 6.3. It can be seen that the SDSM model shows a good agreement between the observed and computed mean daily rainfall, standard deviation and variance with correlation coefficient of 0.57 and 0.50 during calibration and validation respectively. Unlike temperature, the correlation coefficient for the precipitation series is at a lower side. This may be attributed to considerable variation in precipitation with respect to time and space.

Table 6.3 Comparison between daily precipitation (Observed) and daily precipitation (Computed) during model calibration and validation

Type	Period	Mean	SD	Var	Correlation, r
Model calibration	Precp_61-91_Observed	3.02	11.70	136.88	0.57
	Precp_61-91_Computed	3.59	7.24	52.47	
Model validation	Precp_92-01_Observed	3.31	14.46	209.01	0.50
	Precp_92-01_Computed	3.44	6.69	44.81	

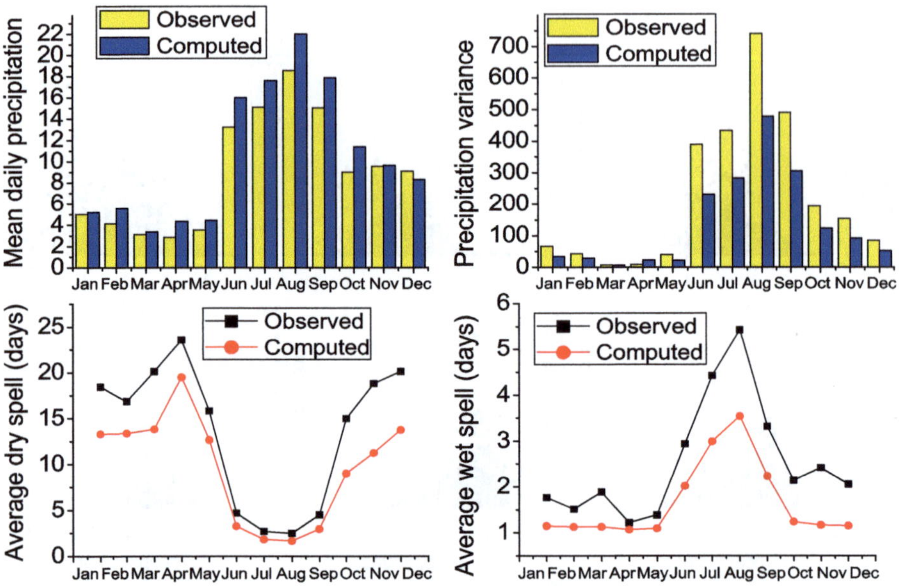

Fig. 6.3 Calibration output of SDSM model downscaling (1961–1991) for daily precipitation

Figure 6.3 highlights the calibration result of the SDSM model with good agreement between observed and computed mean daily precipitation. The computed variance is greater for the monsoon months (June to September). However, there is under-estimation of average dry and wet-spell length. The validation result of the SDSM model for the period 1992–2001 between observed and computed mean daily precipitation, variance, dry-spell length and wet-spell length is shown in Fig. 6.4.

6.5.3 Future Scenario Generation

The validated Multiple Linear Regression models between the predictand and large-scale predictors are used to generate the future downscaled data using the HadCM3 GCM data for A2 scenario. The result of the downscaled daily rainfall for different periods is shown in Fig. 6.5. The figure clearly indicates an increasing rainfall trend in the corresponding months for different periods.

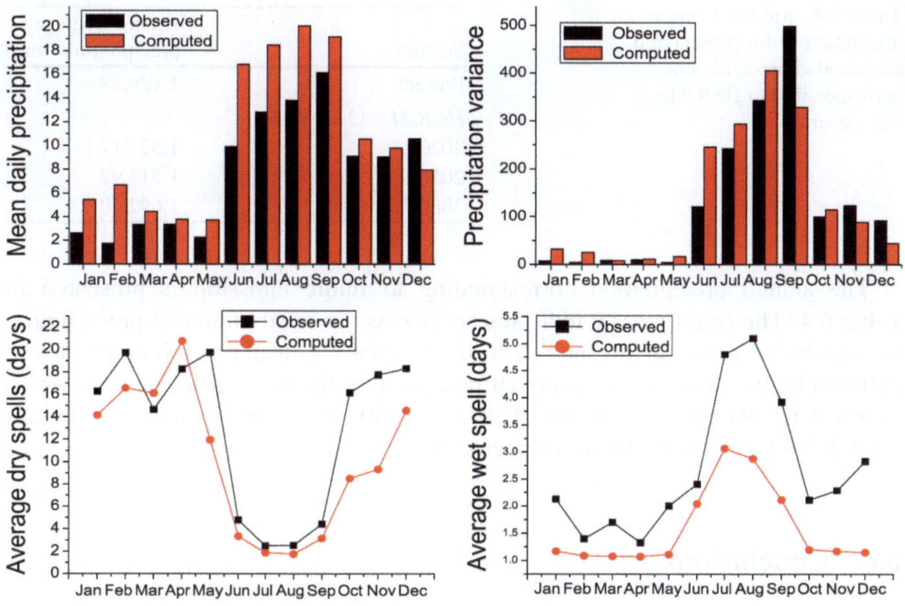

Fig. 6.4 Validation output of SDSM model downscaling (1992–2001) for daily precipitation

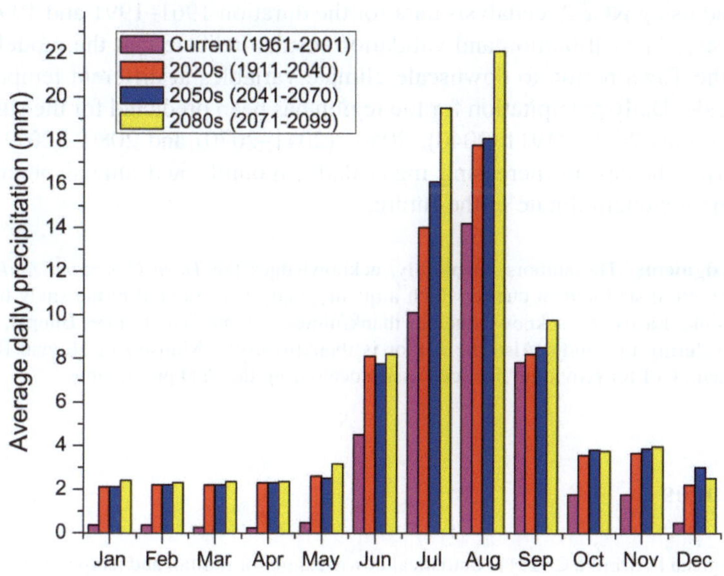

Fig. 6.5 General trend of mean daily precipitation corresponding different scenarios

Table 6.4 Annual average precipitations for present and downscaled precipitation corresponding to HadCM3 A2 scenario

Scenario	Annual average precipitation (mm)
Present	1,129.58
HadCM3 A2 scenario	
2020s	1,327.17
2050s	1,515.92
2080s	1,692.70

The annual precipitation corresponding to future emission is presented in Table 6.4. The result clearly indicates an increase in trend of annual precipitation for successive scenarios. In the 2020s, the simulated annual precipitation is about 200 mm higher than the mean annual precipitation for the present scenario which stands at 1,129 mm. Similarly for 2050s and 2080s, the annual mean precipitations are 1,515.92 and 1,692.70 mm respectively.

6.6 Conclusions

SDSM is one of the downscaling tools widely used to downscale simulated GCM data into local fine-scale data. In the present study, multiple linear regression based SDSM model has been used to downscale daily precipitation data corresponding to HadCM3 A2 GCM (1961–2099). The model calibration and validation has been performed using NCEP reanalysis data for the duration 1961–1991 and 1992–2001 respectively. The calibration and validation results indicate that the model can be used in the Tawa region to downscale climate variables at different temporal and spatial scale. Daily precipitation for the region has been predicted for the study area for the periods 2020s (1911–2040), 2050s (2041–2070) and 2080s (2071–2099). The study indicates an increasing mean daily, monthly and annual precipitation suggesting a wetter climate in the future.

Acknowledgments The authors, thankfully, acknowledges the *Tawa Command Authority* for providing rainfall station data due to which a quality analysis of rainfall in the study have been made possible. Lastly, we acknowledge our thankfulness to State Data Centre, Bhopal, India for supporting during the study. Also the author is thankful to the Ministry of Human Resources Development, GoI for providing financial assistance during the PhD programme.

References

Aksornsingchai P, Srinilta C (2011) Statistical downscaling for rainfall and temperature prediction in Thailand. In: Proceedings of the international multiconference of engineers and computer scientists, vol I. Hong Kong, March 16–18

Bardossy A (1997) Downscaling from GCM to local climate through stochastic linkages. J Environ Manage 49:7–17

Crawford T, Bretts NS, Favis-Mortlock D (2007) GCM grid-box choice and predictor selection associated with statistical downscaling of daily precipitation over Northern Ireland. Clim Res 34:145–160

Ghosh S, Mujumdar PP (2006) Future rainfall scenario over Orissa with GCM projections by statistical downscaling. Curr Sci 90(3):396–404

Ghosh S (2010) SVM-PGSL coupled approach for statistical downscaling to predict rainfall from GCM output. J Geophys Res 115, D22102. doi:10.1029/2009JD013548

Hashmi MZ, Shamseldin AY (2011) Comparison of SDSM and LARS-WG for simulation and downscaling of extreme precipitation events in a watershed. Stoch Environ Res Risk Assess 25:475–484

Hassan Z, Harun S (2012) Application of statistical downscaling model for long lead rainfall prediction in Kurau river catchment of Malaysia. Malays J Civ Eng 24(1):1–12

Hewitson BC, Crane RG (1996) Climate downscaling: techniques and application. Climatic Res 7:85–95

Huth R (2002) Statistical downscaling of daily temperature in Central Europe. J Climate 15:1731–1742

Kannan S, Ghosh S (2010) Prediction of daily rainfall state in a river basin using statistical downscaling from GCM output. Stoch Environ Res Risk Assess. doi:10.1007/s00477-010-0415-y

Kodra E, Ghosh S, Ganguly AR (2012) Evaluation of global climate models for Indian monsoon climatology. Environ Res Lett. doi:10.1088/1748-9326/7/1/014012

Lall M, Nozawa T, Emori S, Harasawa H, Takahashi K, Kimoto M, Abe-Ouchi A, Nakajima T, Takemura T, Numaguti A (2001) Future climate change: implications for Indian summer monsoon and its variability. Curr Sci 81(9):1196–1207

Murphy J (2000) Predictions of climate change over Europe using statistical and dynamical downscaling techniques. Int J Climatol 20:489–501

Ojha CSP, Goyal MK, Adeloye AJ (2010) Downscaling of precipitation for lake catchment in arid region in India using multiple linear regression and neural networks. Open Hydrol J 4:122–136

Prudhomme C, Jakob D, Svensson C (2003) Uncertainty and climate change impact on the flood regime of small UK catchments. J Hydrol 277:1–23

Raje D, Mujumdar PP (2009) A conditional random field-based downscaling method for assessment of climate change impact on multisite daily precipitation in the Mahanadi basin. Water Resour Res 45:1–20

Raje D, Mujumdar PP (2011) A comparison of three methods for downscaling daily precipitation in the Punjab region. Hydrol Process 25:3575–3589

Robock A, Turco RP, Harwell MA, Ackerman TP, Andressen R, Chang HS, Sivakumar MVK (1993) Use of general circulation model output in the creation of climate change scenarios for impact analysis. Clim Change 23:293–335

Wilby RL, Wigley TML (1997) Downscaling general circulation model output: a review of methods and limitations. Progr Phys Geogr 21:530–548

Wilby RL, Dawson CW (2007) SDSM 4.2 - a decision support tool for the assessment of regional climate change impacts. UK

Chapter 7
Drought Monitoring of Chhattisgarh Using Different Indices Based on Remote Sensing Data

T. Mukherjee, S. Mukherjee, A. Mukhopadhaya, A.K. Roy, and S. Dutta

Abstract The mid-latitude and new born state Chhattisgarh faces drought due to poor and delayed monsoon, very high summer temperature and in sufficient water resources. In the present study detailed analysis of vegetative and meteorological drought indicative indices of the study area, has been carried out for the years 2009–2012. Standardised Precipitation Index (SPI) has been used to quantify the precipitation deficit. It is calculated using TRMM Monthly gridded rainfall data. Vegetative drought indices like Vegetation Health Index (VHI) obtained from Global Vegetation Index (GVI) of NOAA AVHRR satellite. Normalized Difference Vegetation Index (NDVI) obtained from MODIS. Detailed analyses of spatial and temporal drought Indices during monsoon and non-monsoon seasons have been carried out through. Drought Index Maps generated in Geographic Information Systems (GIS) environment. Analysis and interpretation of these maps reveal that there is severe drought condition in Chhattisgarh on 2009, 2011 and 2012. Mainly pre-monsoon condition have faced drought due to lack of rainfall. So, Drought and precipitation are interdependent. There is good correlation between SPI and VHI, and Poor correlation between SPI and NDVI. Also there is some correlation between VHI and NDVI. These indices are also inter-dependent, interrelated and good indicator of drought.

Keywords Meteorological drought • Normalized Difference Vegetation Index • Standardised Precipitation Index • Vegetation Health Index • Vegetative drought

---

T. Mukherjee • S. Mukherjee (✉)
Department of Natural Resources, TERI University, New Delhi 110070, India
e-mail: sandip.iirs@gmail.com

A. Mukhopadhaya
School of Oceanographic Studies, Jadavpur University, Kolkata, India

A.K. Roy • S. Dutta
N.P.G.C, University of Lucknow, Lucknow 226001, India

M. Singh et al. (eds.), *Climate Change and Biodiversity: Proceedings of IGU Rohtak Conference, Vol. 1*, Advances in Geographical and Environmental Sciences,
DOI 10.1007/978-4-431-54838-6_7, © Springer Japan 2014

7.1 Introduction

Drought is a general climatic phenomenon which affects people most. In India drought is very common for most of the states. Research has shown that the lack of a precise and objective definition in specific situations has been an obstacle to understanding drought which has led to indecision and inaction on the part of managers, policy makers, and others (Wilhite and Glantz 1985; Wilhite et al. 1986). Nearly 50 % of the world's most populated areas are highly vulnerable to drought and more importantly, almost all of the major agricultural lands are located there (USDA 1994). In India, these drought areas are mainly confined to the Peninsular and Western parts of the country. Out of 3.28 million km^2 of geographical area in India about 1.07 million km^2 of land are subjected to different degrees of water stress and drought conditions (Mishra et al. 2005).

There are mainly three types of drought which has seen in India, i.e. Agricultural drought, Hydrological drought and Meteorological drought. All of them have adverse effect in Socio-economic, Agricultural and Environmental conditions. There is various methods and indices, developed by many scientists (e.g. Palmer 1965, 1968; Gibbs and Maher 1967; Shafer and Dezman 1982; Kogan 1990, 2002; McKee et al. 1993; Keyantash and Dracup 2004) for drought monitoring and analysing, using Remote Sensing technique. They have taken some bio-physical parameter into consideration such as rainfall, soil moisture, potential evapo-transpiration, vegetation condition, ground-water and surface-water levels. They are drought responsive as well as drought causative parameters and also bear some correlation with each other. They can be used for drought monitoring, but requirement varies from place to place. In Indian Condition, Asian South-west monsoon plays a very crucial role in irrigation and ground water recharge for Chhattisgarh. And also there is no perennial river system in this region. So, water scarcity is the major reason for drought. Uncertainty of onset of monsoon causes mild to severe drought in Chhattisgarh. This meteorological drought turn affect the growth of the natural vegetation and crop production, which results Agricultural drought. Drought condition is not so much frequent in Chhattisgarh. It receives 1,292.1 mm rainfall per year and 61 rainy days (Chhattisgarh at a glance 2002). But due to some other parameters like soil moisture capacity, soil type, slope, annual temperature etc. the amount of ground recharge is poor.

In this study, detailed analysis of annual drought monitoring has been carried out to identify spatio-temporal drought patterns in Agricultural or vegetative drought and meteorological drought. The final output is generated to show relation among the indices. Standardised Precipitation Index (SPI) has been used to monitor meteorological drought. Vegetation Health Index, Normalized Difference vegetation Index has been employed to assess Agricultural drought. Finally all the outputs are correlated with each other.

7.2 Study Area

Chhattisgarh, a twenty-first century State, came into being on November 1, 2000, situated at the heart of India (Extension: 17°46′N to 24°5′N & 80°15′E to 84°20′E) (Fig. 7.1). This state is the 10th largest state of India with an area of more than 135,000 km^2. It ranks as 16th among most populated state of India. The population of Chhattisgarh is 25,540,196 & density is 189 person/km^2 (2011 Census Report). One third of Chhattisgarh's population is of tribes, mostly in the thickly forested areas in the North and South. It comes under the hot Torrid Zone and the state observes tropical type of climate. Though weather varies from region to region, it's warm in most of the places. Late in the month of June, South-West Monsoons (July–September) arrive in the state. Chhattisgarh receives decent amount of rainfall with an average of 1,292 mm. Since it falls under the rice-agro-climatic zone, rainfall is the main source of irrigation. A significant variation in the annual rainfall affects the harvest and results drought condition (Chhattisgarh State Report).

7.3 Materials

7.3.1 Satellite Data Used

Three indices have used for this study- Vegetation health Index (VHI), Normalized Difference Vegetation Index (NDVI), and Standard Precipitation Index (SPI) (Table 7.1).

7.3.2 Vegetation Health Index

For this study, VHI products have downloaded from AVHRR Global Vegetation Index-X (GVI-X) Data (http://www.star.nesdis.noaa.gov/smcd/emb/vci/VH/vh_ftp.php). The GVI-x processing system comprises of a single software application, written in the C programming language primarily for the GNU/Linux operating system. It acts as a data aggregator, ingesting data for one NOAA satellite from the Clouds from AVHRR (CLAVR-x) processing system. GVI-x output is a subset of orbital AVHRR data extracted from the CLAVR-x output and collated onto a global GVI-x grid. GVI-x output is a single file for each processing period in the Hierarchical Data Format (HDF) (GVI-x Vegetation Health Product User Guide, Version 0.2, Sept. 2, 2008.) To identify Vegetative and agricultural droughts, vegetation condition and temperature condition are taken as two major parameters and they

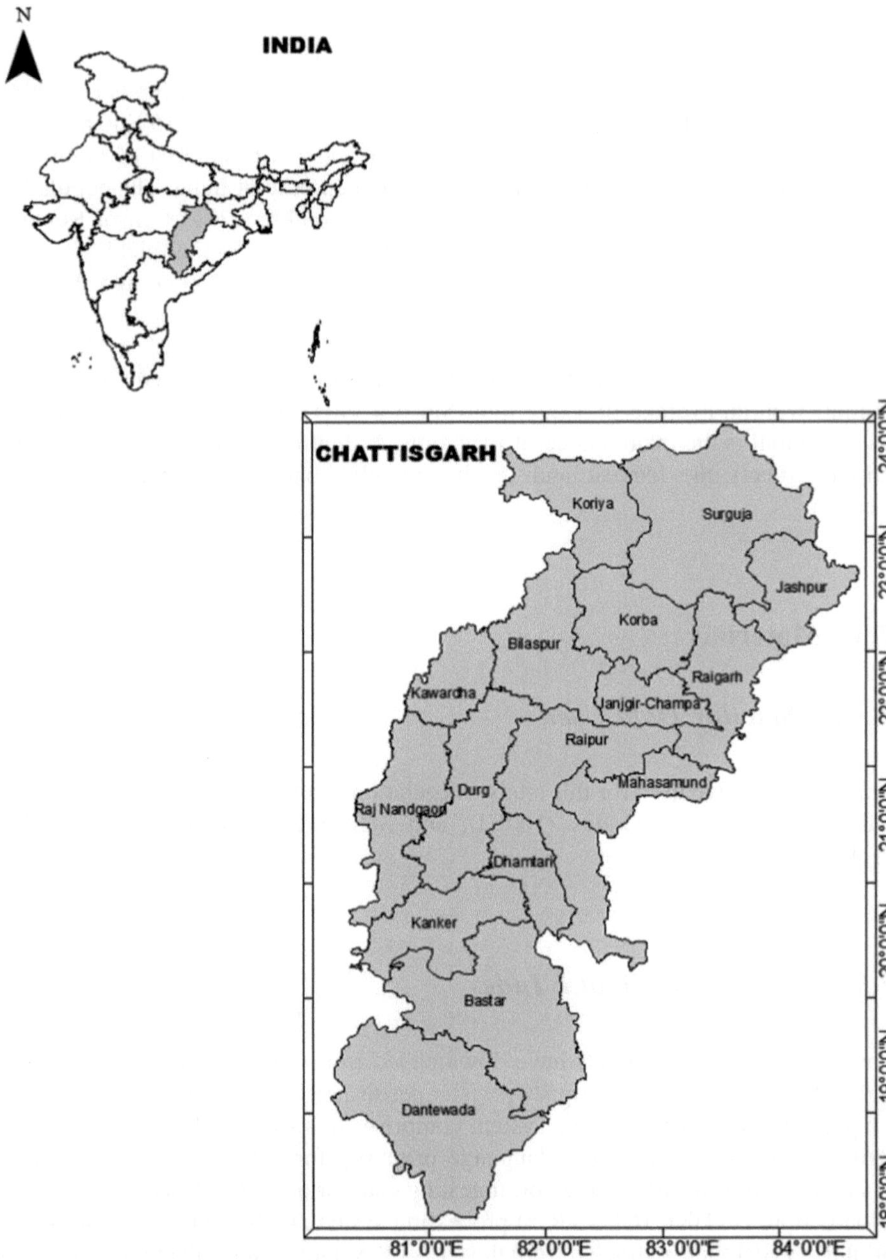

Fig. 7.1 Location map of study area

Table 7.1 Description of satellite data product used

			Resolution	
Data product used	Satellite (sensor)	Data format	Spatial	Temporal
Vegetation Health Index (VHI)	NOAA-AVHRR	HDF (.hdf)	16 KM	8 day
Normalized Vegetation Difference Index (NDVI)	MODIS	HDF (.hdf)	1 KM	30 day
Rainfall	TRMM	ASCII	0.25°×0.25° Grid	30 day

reflect vegetation-stress. Vegetative Condition Index is related to the long-term minimum and maximum NDVI (Kogan 1990) and is related as:

$$NDVI = 100 \times \frac{NIR - R}{NIR + R} \tag{7.1}$$

$$VCI = 100 \times \frac{NDVI_{MAX} - NDVI}{NDVI_{MAX} - NDVI_{MIN}} \tag{7.2}$$

where NDVI, $NDVI_{min}$, and $NDVI_{max}$ are the seasonal average NDVI, its multi-year absolute minimum and its maximum, respectively.

The TCI (Kogan 1990) is given as:

$$TCI = 100 \times \frac{BT_{MAX} - BT}{BT_{MAX} - BT_{MIN}} \tag{7.3}$$

where BT, BT_{min}, and BT_{max} are the seasonal average of weekly brightness temperature, its absolute minimum, and maximum, respectively.

VCI and TCI characterise by varying moisture and thermal conditions of vegetation, Vegetation Health Index represents overall vegetation health which was used by Kogan (2001) who also gave five classes of VHI that was used for drought mapping. VHI is computed and expressed as:

$$VHI = 0.5(VCI) + 0.5(TCI) \tag{7.4}$$

7.3.3 Normalized Difference Vegetation Index

NDVI is the first derived product from MODIS. It is derived using NOAA-AVHRR dataset. It has done for last 27 year (1981–2009).

$$NDVI = \frac{NIR - R}{NIR + R} \tag{7.5}$$

Table 7.2 Distribution of minimum, maximum and mean rainfall (in mm) of pre-monsoon and post-monsoon from 2009 to 2012

Months	Year	2009	2010	2011	2012
June	Min	0.834	7.184	37.773	21.923
	Mean	86.99232	141.7598	293.1442	155.6784
	Max	444.108	510.48	946.02	655.83
July	Min	112.348	201	117.09	138.75
	Mean	444.4509	535.845	305.2279	417.4288
	Max	863.906	890	544.039	973.11
August	Min	95.597	96.971	179	97.2
	Mean	295.2701	309.3534	502.4733	410.9432
	Max	633.971	896.804	989	669.828
September	Min	56.837	105.36	26.415	140.86
	Mean	201.6091	257.6074	294.7414	263.073
	Max	538.152	472.255	719.13	524.188
October	Min	0.739	14.095	0	0.388
	Mean	87.12977	185.8339	32.37924	41.38416
	Max	542.61	860.85	476.099	213.221

7.3.4 *Rainfall Data*

Here Tropical Rainfall Measuring Mission (TRMM) Monthly 0.25 × 0.25 grid Rainfall data had been used. TRMM is a joint venture of USA and Japan. The rainfall data was in ASCII format (Table 7.2).

7.3.5 *Methodology*

7.3.5.1 Image Processing and Calculation

VHI Series dataset (2009–2012) were imported in tiff format. The study areas were subsetted using the AOI of study area. All the data sets are processed and georeferenced. Then Mean Images were calculated using model builder to make monthly mean VHI images. Then all the images were multiplied with scale factor, by default it is 0.00. All the images were co-registered with the shapefile of study area. Then the study areas for June to October months of each year (2009–2012) were extracted using shapefile.

NDVI datasets for June to October Months for 4 years i.e. 2009–2012 were in .hdf format. Only 1 KM NDVI products were imported using ERDAS Imagine. These images were georeferenced, but these were reprojected to bring all the datasets in same projection and datum. After that Mosaicing was done and the study area was extracted using AOI.

Accumulated Rainfall gridded datasets from TRMM were in ASCII format. Previously all of the datasets for 2009–2012 were saved in a text file to import them in a Excel sheet. After converting the data into excel format, point vector layer for rainfall has created using ArcGIS. Then using Inverse Distance Weighted (IDW) Method interpolated surface has created with the help of point vector data. This is done for 2009 (June–October), 2010 (June–October), 2011 (June–October) and 2012 (June–October). All these datasets have used to calculate SPI (Fig. 7.2).

SPI is based on the probability of precipitation for any desired time scale. The SPI is based on the probability of precipitation for any desired time scale and spatially invariant indicator of drought (Guttman 1998, 1999). This Index has proposed by McKee et al. (1993, 1995). It is used to weigh up irregular and intense precipitation. It involves fitting a gamma probability density function to a given frequency distribution of precipitation totals for a station (Edwards and McKee 1997). The SPI is computed by dividing the difference between the normalised seasonal precipitation and its long-term seasonal mean by the standard deviation. Thus,

$$SPI = \frac{X_{ij} - X_{im}}{s} \tag{7.6}$$

where, X_{ij} is the seasonal precipitation at the ith rain gauge station and jth observation, X_{im} the long-term seasonal mean and s is its standard deviation. Interpolated Surface of SPI datasets are used in the present study.

7.3.5.2 Drought Zonation and Analysis

All the three datasets (e.g. VHI, NDVI & SPI) are used to monitor effective drought condition. For good understanding of drought conditions all the indices were divided into some drought zones which will depict the intensity of drought viz. Extreme drought, Severe drought, Moderate Drought, Mild Drought and No Drought. The Zonation has done on the basis of literatures and previous researches on this field (Bhuiyan et al. 2006; Kogan 2002; Mckee et al. 1993; Karinielli et al. 2009). According these there are two terminologies which are used to describe the drought situation, i.e. Monsoon Period & Non-Monsoon Period (Bhuiyan 2004) The Non-Monsoon Period is further divided into two periods, i.e. Pre-monsoon and Post-monsoon. All these climatic conditions are equally important for drought mapping. For Drought Mapping following Classification scheme have used in this study (Fig. 7.3).

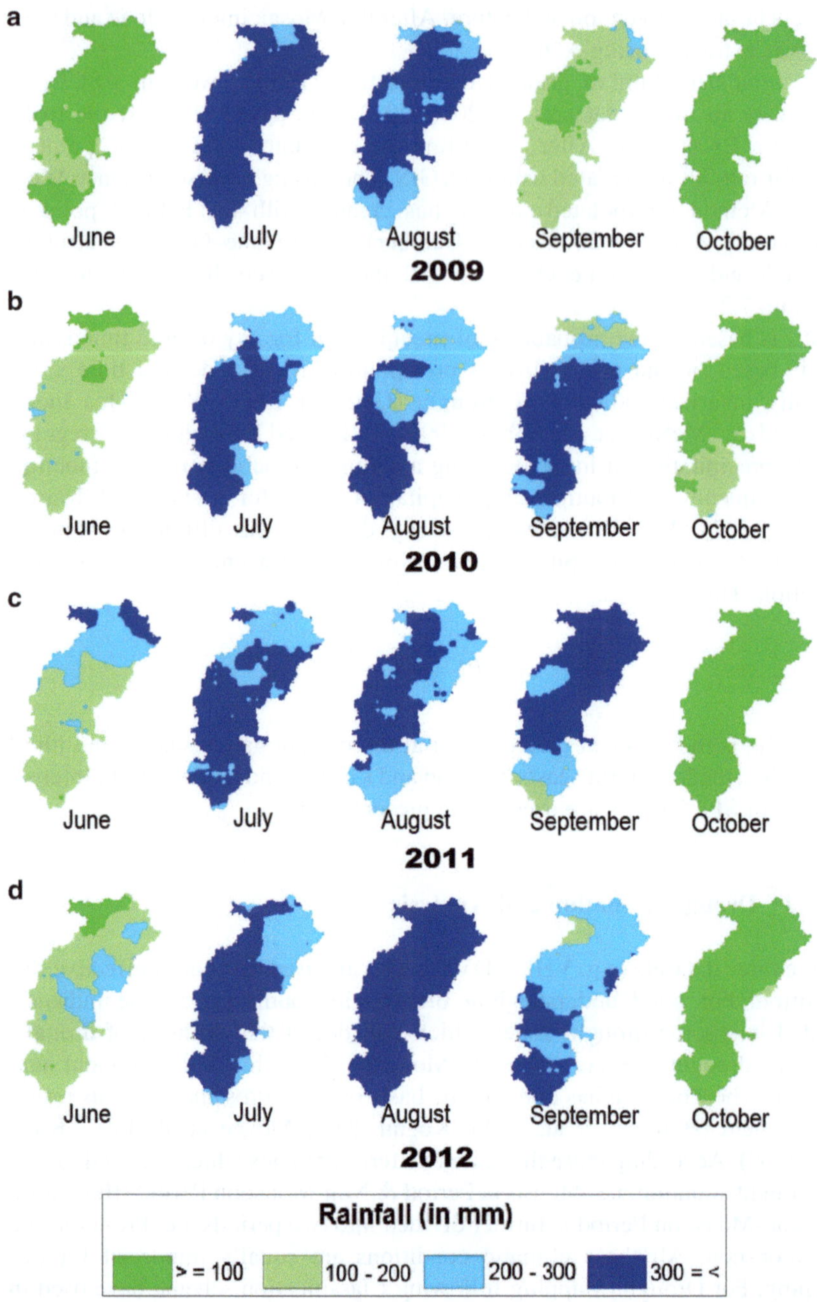

Fig. 7.2 Showing rainfall distribution for year (**a**) 2009, (**b**) 2010, (**c**) 2011, (**d**) 2012

Drought Classes \ Classification Scheme	VHI (Kogan, 2002)	SPI (McKee et al., 1993)	NDVI (Karinieli et al., 2009)
Extreme Drought	< 10	< - 2.0	< 0.02
Severe Drought	10 - 20	< - 1.5	
Moderate Drought	20 - 30	< - 1.0	0.02 - 0.05
Mild Drought	30 - 40	< - 0.0	
No Drought	> 40	> - 0.0	> 0.05

Fig. 7.3 Classification scheme for drought mapping

7.4 Results and Discussion

7.4.1 Vegetative Drought

Mean of the VHI images of the study area have been computed for the monsoon and non-monsoon periods (June–October) during the years 2009–2012 by averaging weekly values. VHI maps have been generated with 16 km spatial resolution, and have been classified to represent various drought intensities (Classification scheme).

From this study it is evident that, vegetation experienced stress and loss of vegetation health in pre-monsoon period (June). During 2009, 2011 & 2012, extreme to severe drought have seen in the month of June. Also moderate to mild drought have seen in Post monsoon period (October). On the other hand, there is good vegetation growth have experienced in the year 2010. And also monsoon months (July–September) have been drought free (Figs. 7.4 and 7.5).

From the VHI classified map the area under various drought conditions has been plotted.

From these graphs we may say that, moderate to mild drought conditions have shown in maximum area, while extreme drought conditions are rare. During the month of June, drought conditions are mostly experienced, while August and September has faced no drought condition.

Also vegetative drought condition has monitored using Normalized Vegetation Index (NDVI). Using MODIS 13A2 Product (Monthly NDVI product) of 1 KM

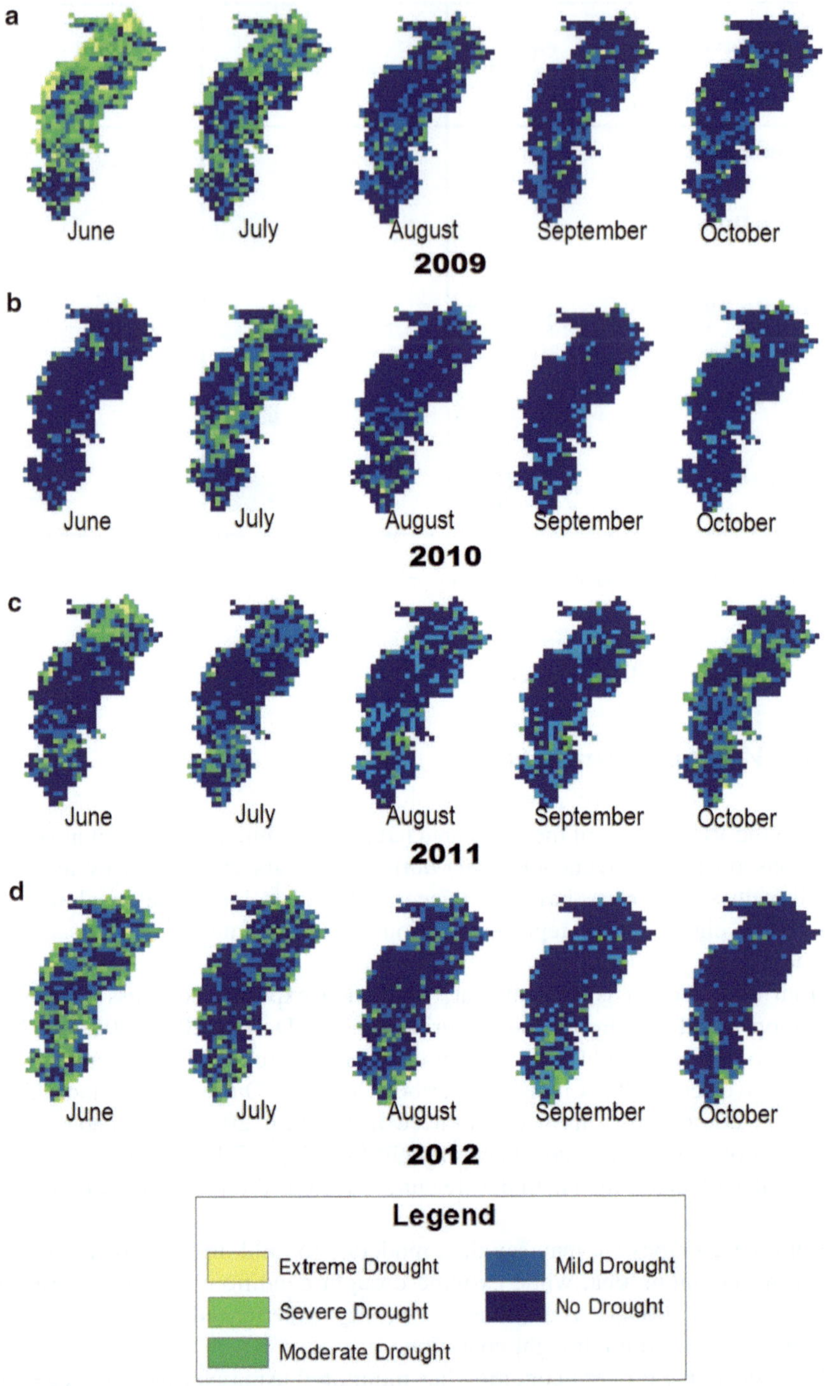

Fig. 7.4 Showing vegetative drought severity for year (**a**) 2009, (**b**) 2010, (**c**) 2011, (**d**) 2012

spatial resolution, this study has done. NDVI is also classified using classification scheme (Fig. 7.6).

As it has discussed earlier, that drought conditions can be experienced in the pre-monsoon period. It have also evidenced by the classified NDVI images. During the month of August, September & October, good vegetation coverage has shown in the area. It proves that there is no drought condition at that time (Post monsoon Period). On the other hand June and July month experience least amount of vegetation cover as expected.

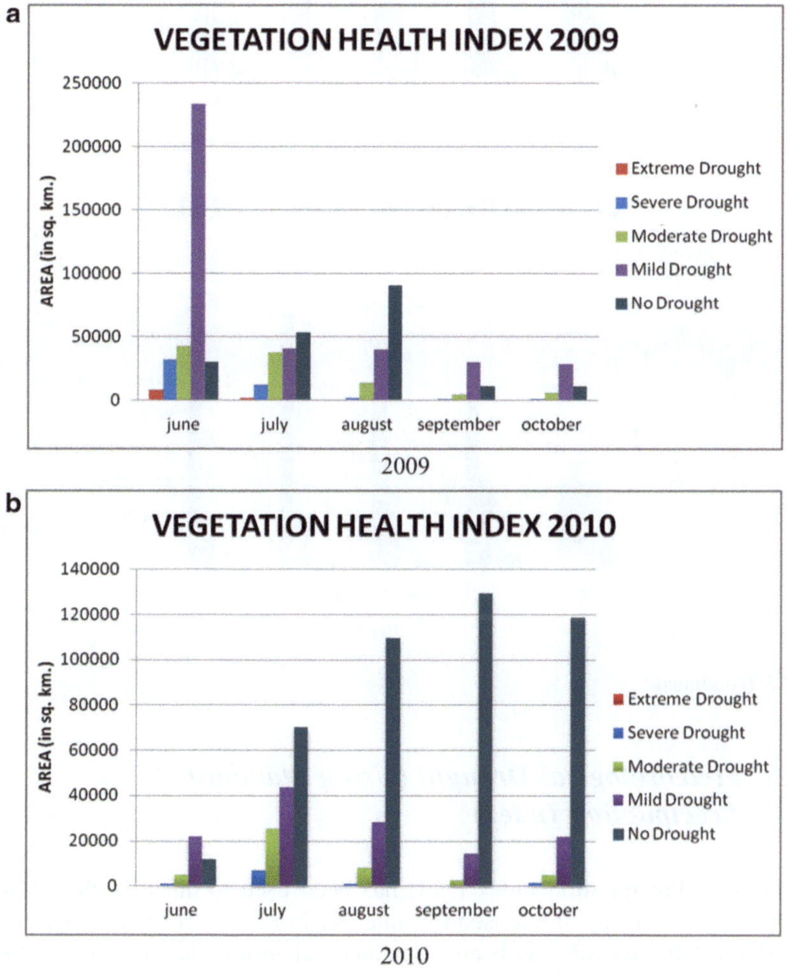

Fig. 7.5 Showing graphs depicting Annual Vegetation health index (**a**) 2009, (**b**) 2010, (**c**) 2011, (**d**) 2012

c

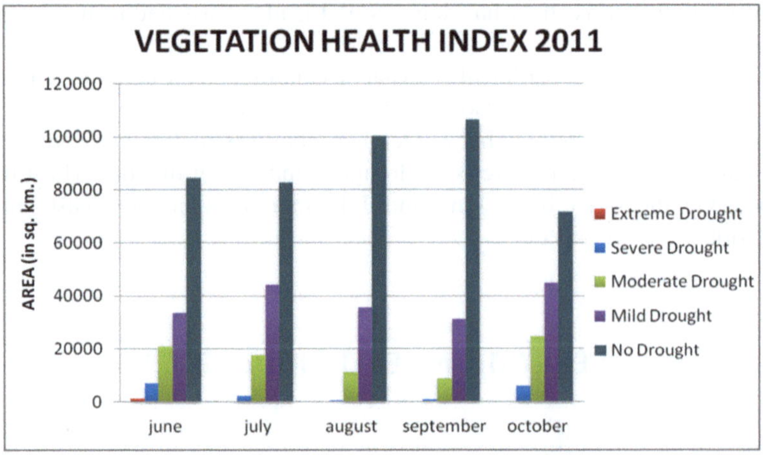

2011

d

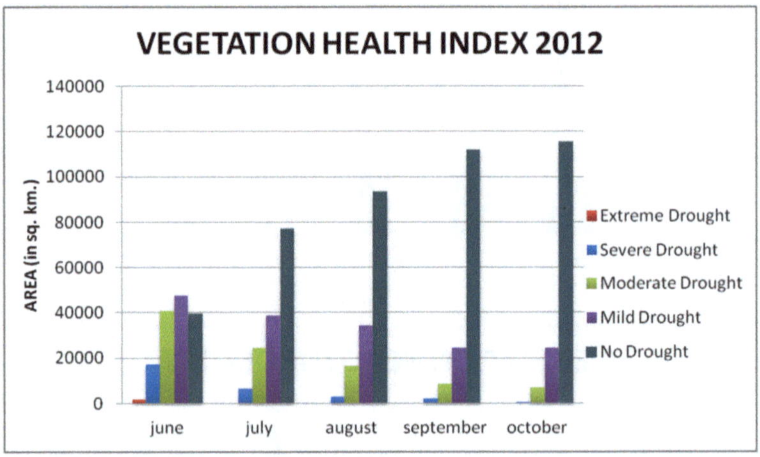

2012

Fig. 7.5 (continued)

7.4.2 Meteorological Drought (Using Standard Precipitation Index)

Standardized Precipitation Index (SPI) has been used to quantify the amount of precipitation deficiency. It has used for June'2009 to July'2012 time span. From the gridded rainfall data SPI has been calculated. Then the classification of SPI has been carried out using the method, proposed by Mc Kee et al. (1995) and explained by Edwards and McKee (1997), to represent various hydro-meteorological drought intensities.

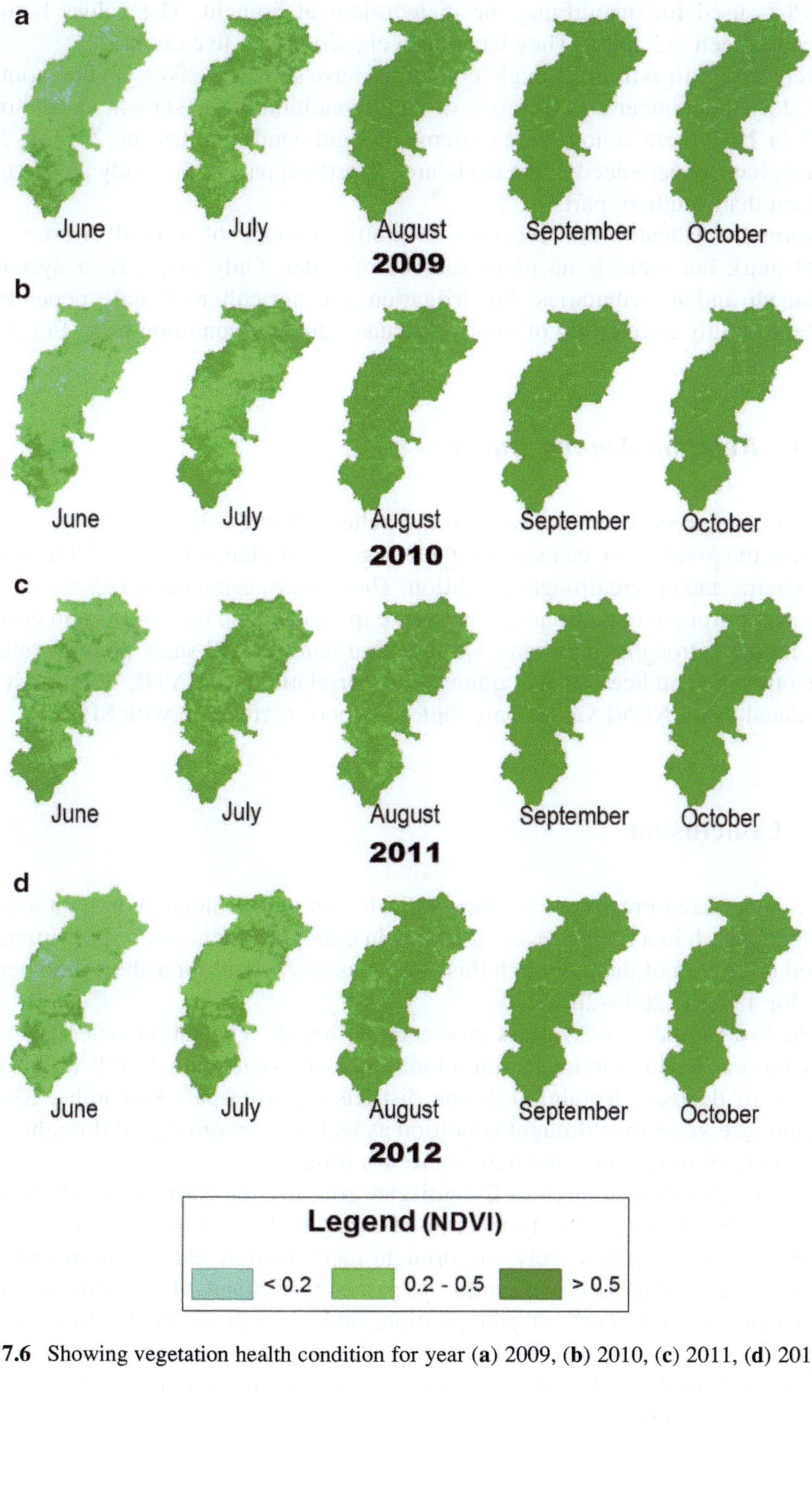

Fig. 7.6 Showing vegetation health condition for year (**a**) 2009, (**b**) 2010, (**c**) 2011, (**d**) 2012

SPI is used for quantifying the meteorological drought. The values between varies between −2 and 2. They have been classified into five classes.

Here severe to extreme drought conditions have shown in 2009. On the month of June, July, September and October, drought conditions have shown. On the other hand, in 2010 there is no severe extreme drought condition. During 2011 & 2012 drought have experienced in the study area. Northern part of the study area is more affected than southern part.

Normally Chhattisgarh receives a pretty amount of rainfall (more than 1,200 mm), but there is no other sources of water. Only major river system is Mahanadi and its tributaries. So, irrigation and agriculture mainly depends on rainfall. Mainly uncertainty of monsoon causes drought condition here (Fig. 7.7).

7.4.3 Relation Among Indices

There is some correlation between all the indices (Table 7.3).

From the p-value, we can say that there is good correlation between VHI and SPI for extreme and severe drought condition. That means, using these indices extreme and severe drought conditions can be well explained. Also they bear good correlation in others drought conditions. On the other hand, NDVI show poor correlation with other two indices. It has quite good correlation with VHI, because it has calculated using NDVI values only, but, very poor correlation with SPI.

7.5 Conclusion

The Standardized Precipitation Maps indicate that meteorological drought appears in Chhattisgarh in a random way. It comes in either pre-monsoon or post-monsoon period or in both of them, though this study has carried out for a short time period, only for 4 years (2009–2012).

The low rainfall causes stress in vegetation health. Vegetation generally maintains normal health for longer duration. But here vegetation health is in stress because of decrease in rainfall. Some districts e.g. Bilaspur, Kawardha, Koriya, Surguja face vegetative drought condition as well as meteorological drought. Here uncertainty of monsoon is the main cause for drought.

The spatio-temporal maps of Chhattisgarh reveal some correlation between SPI and VHI. Furthermore, the time limit of drought development and duration of drought also varies extensively. So, drought identification, classification and analysis are highly influenced by monitoring parameters. Standard Precipitation Index (SPI) monitors deficiency in precipitation, and Vegetation Health Index (VHI) monitors vegetation health. VHI represent the negative impact of poor meteorological and hydrological conditions on water and vegetation, as a result VHI present better picture of drought.

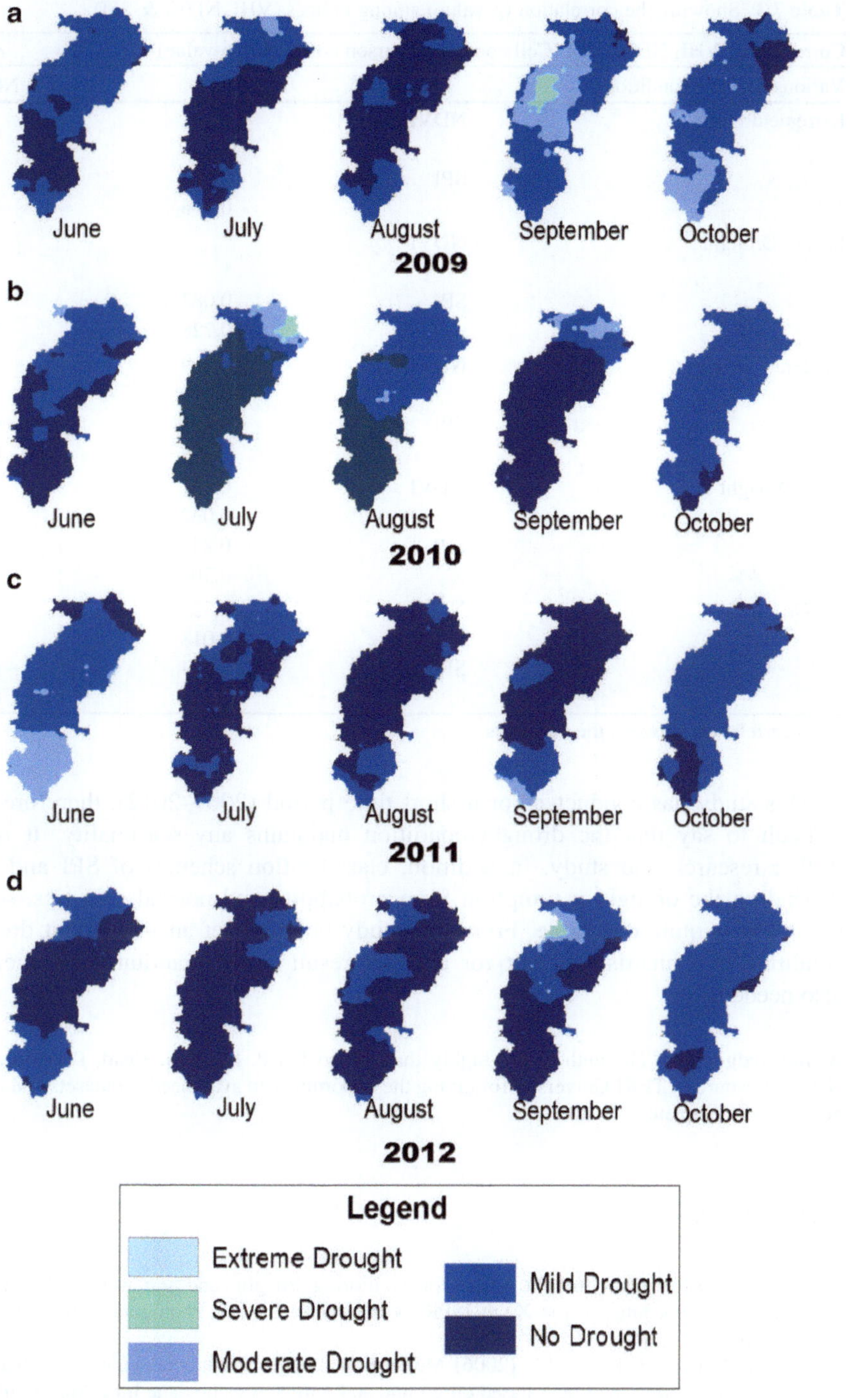

Fig. 7.7 Showing meteorological drought using SPI for year (**a**) 2009, (**b**) 2010, (**c**) 2011, (**d**) 2012

Table 7.3 Showing the correlation (p-value) among indices (VHI, NDVI & SPI)

Correlations: VHI, NDVI, SPI (Cell contents: Pearson correlation p-value)			
Various drought conditions		VHI	NDVI
Extreme drought	NDVI	*	
		*	
	SPI	−0.516	*
		0.816	*
Severe Drought	NDVI	*	
		*	
	SPI	−0.083	*
		0.728	*
Moderate Drought	NDVI	0.519	
		0.019	
	SPI	−0.199	−0.200
		0.399	0.397
Mild Drought	NDVI	0.481	
		0.032	
	SPI	0.241	0.104
		0.306	0.662
No Drought	NDVI	0.525	
		0.018	
	SPI	0.094	−0.269
		0.728	0.251

* No correlation between the variables

This study has conducted for a short time period (2009–2012), therefore it is difficult to say that the drought condition maintains any seasonality. It needs further research and study. In addition, classification schemes of SPI and VHI depend on the drought assumption from probability and normal statistics, so the reliability is quite debatable. From this study one can get an idea about drought condition of Chhattisgarh, but for a better result some quantitative application also needed.

Acknowledgement The authors are highly thankful to Dr. P. K. Joshi, Head, Department of Natural Resources, TERI University for giving the opportunity to avail needed datasets and useful guidance to complete this study.

References

Bhuiyan C (2004) Various drought indices for monitoring drought condition in Aravalli terrain of India. In: Proceedings of the XXth ISPRS Conference. Int Soc Photogramm Remote Sens, Istanbul

Bhuiyan C, Singh RP, Kogan FN (2006) Monitoring drought dynamics in the Aravalli region (India) using different indices based on ground and remote sensing data. Int J Appl Earth Obs Geoinf 8:289

Edwards DC, McKee TB (1997) Characteristics of 20th century drought in the United States at multiple timescales, Climatology Report No 97–2. Colorado State University, Fort Collins
Gibbs WJ, Maher JV (1967) Rainfall deciles as drought indicators. Bureau of Meteorology Bulletin No 48, Commonwealth of Australia, Melbourne
Guttman NB (1998) Comparing the palmer drought index and standardized precipitation index. J Am Water Resour Assoc 34(1):113–121
Human Development Report of Chhattisgarh (2005) Government of Chhattisgarh, Chhattisgarh
Jelenak A (2004) Experimental global vegetation index processing system: GVI-x version 0.05, NOAA
Karnieli A, Sprintsin M, Berliner P, Rotenberg E, Yakir D, Cohen S (2009) Evaluating the performance of the MODIS Leaf Area Index (LAI) product over a Mediterranean dryland planted forest. Int J Remote Sens 30(19):5061–5069
Keyantash J, Dracup JA (2004) An aggregate drought index: assessing drought severity based on fluctuations in the hydrologic cycle and surface water storage. Water Resour Res 40. doi:10.1029/2003WR002610
Kogan FN (1990) Remote sensing of weather impacts on vegetation in non-homogeneous areas. Int J Remote Sens 11(8):1405–1419
Kogan FN (2001) Operational space technology for global vegetation assessment. Bull Am Meteorol Soc 82(9):1949–1964
Kogan FN (2002) World droughts in the new millennium from AVHRR-based Vegetation Health Indices. Eos Trans Am Geophy Union 83(48):562–563
McKee TB, Doesken NJ, Kleist J (1993). The relation of drought frequency and duration to time scales. Proceedings of the eighth conference on applied climatology. Am Meteorol Soc, Boston, pp 179–184
McKee TB, Doesken NJ, Kleist J (1995) Drought monitoring with multiple time scales. In: Proceedings of the ninth conference on applied climatology, Am Meteorol Soc Boston, pp 233–236
Palmer WC (1965) Meteorological drought. Research Paper No. 45, U.S. Department of Commerce Weather Bureau, Washington, DC
Palmer WC (1968) Keeping track of crop moisture conditions, nationwide: the new crop moisture index. Weatherwise 21:156–161
Shafer BA, Dezman LE (1982) Development of a Surface Water Supply Index (SWSI) to assess the severity of drought conditions in snowpack runoff areas. In: Proceedings of the Western Snow conference. Fort Collins, pp 164–175
USDA (1994) Major world crop areas and climatic profiles. World Agricultural Outlook Board, US Department of Agriculture, Agricultural Handbook No 664, pp 157–70
Wilhite DA, Glantz MH (1985) Understanding the drought phenomenon: the role of definitions. Water Int 10:111–120
Wilhite DA, Rosenberg NJ, Glantz MH (1986) Improving federal response to drought. J Clim Appl Meteorol 25:332–342

Chapter 8
Historical Temporal Trends of Climatic Variables Over Kashmir Valley and Discharge Response to Climate Variability in Upper Jhelum Catchment

Rashid A. Wani

Abstract The Himalayan region is one of the most ecologically fragile ecosystems on earth. The global temperature change is likely to have a drastic impact on the hydrology, biodiversity, agriculture and others thereby effecting the people as well as the earth's environment. Temperature data for six instrumental records in the Western Mountains of the Himalayas have been analyzed for seasonal and annual trends over the period of 1975–2009. The records were analyzed by fitting a linear least squares trend line to the annual deviation from the mean and assessing the significance of trend using Student's t test. The impact of observed seasonal temperature and precipitation on discharge is also explored using Kendall's correlation. Strong contrasts are found between the behavior of winter and summer temperatures and between maximum and minimum temperatures. Winter mean and maximum temperature show significant increases while minimum summer temperatures show consistent decline. The precipitation of the area is also shows signs of decline. The study depicts a significant correlation between precipitation and discharge and a weak correlation between temperature and discharge in the Upper Jhelum Basin.

Keywords Correlation • Discharge • Precipitation • Students t test • Temperature

8.1 Introduction

Throughout the earth's history climate has fluctuated between periods of relative warmth and cold climatic variation has a large impact on mans activities. Since climate change will have profound and possible harmful impact on life on the earth and also since many of these causes can be traced to human activity the study of

---

R.A. Wani (✉)
Government Boys Degree College, Sopore 193201, India
e-mail: rsogami@yahoo.co.in

M. Singh et al. (eds.), *Climate Change and Biodiversity: Proceedings of IGU Rohtak Conference, Vol. 1*, Advances in Geographical and Environmental Sciences,
DOI 10.1007/978-4-431-54838-6_8,

climate has become very important in recent times (Singh and Roy 2002). The Third Assessment Report of the Intergovernmental Panel on climate change in 2001 reported that the global mean surface temperature has risen by 0.6 °C over the twentieth century. According to Jones and Moberg (2003), Surface air temperatures are rising globally. Brown et al. (2008) also indicate that extreme daily maximum and minimum temperatures have warmed for most regions of the world since 1950. Sahsamanoglou and Makrogiannis (1992) have proved that during the period 1950–1988, the air temperature in the region of Western Mediterranean presents positive trend of 0.01–0.02 °C/year and equivalent negative trend of 0.01–0.02 °C/year in the region of Eastern Mediterranean as a result of the small change in circulation observed in the region during the examined period. The valley receives most of the precipitation from the western disturbances which originate in the Meditterranean Sea. Climate change will have environmental and social impacts that will likely increase uncertainty in water supplies and agricultural production for people across Asia (Xu 2009). Tsonis et al. (2007) have shown that synchronous behaviour of the various ocean oscillations can provide an explanation for global and local oscillation in temperature. The globally averaged surface temperature is projected to increase by 1.4 to 5.8 °C over the period 1990–2100 (Intergovernmental Panel on Climate Change (IPCC) 2001). The possibility of climate change through man kind's influence in loading the atmosphere with greenhouse gases, aerosols and other constituents has been widely discussed (IPCC 1995). Trends of temperature aren't uniform over the NWH region and annual temperature has risen by 1.6 °C in the last century with winter warming (Bhutiyani et al. 2007). Increasing trend of temperature in the post monsoon and winter and decreasing trend in monsoon has been observed for Srinagar, Shimla, Musoorie, Dehradun and Mukteshwar whereas Leh has experienced warming trend during the last 100 years (Pant et al. 1998). Tropical land-surface precipitation measurements indicate that precipitation is likely to have increased by about 0.2–0.3 % per decade over the twentieth century, but increases are not evident over the past few decades and the amount of tropical land (versus ocean) area for the latitudes 10°N to 10°S is relatively small.

8.2 Study Area

The study area is located in the northwest corner of India in the state of Jammu and Kashmir. Its geographical position is between 30°17′ N and 37°5′ N latitude and 73°26′ E and 80°30′ E longitude. Kashmir is situated within the Himalayan mountain system. Quazigund and Srinagar lies in the valley whereas all the others lie in the hills surrounding the valley. The valley of Kashmir was formed by folding and faulting as the Himalayan mountain chain was thrust between the Indian sub continent and the rest of Asia. The valley runs northwest to southeast along the strike of the mountain chain and is drained by the river Jhelum which cuts through the Pir Panjal at the Baramullah gap. This structural basin is 135 km in length with a maximum width of 40 km and ranges in altitude from 5,200–6,000 ft above the sea level.

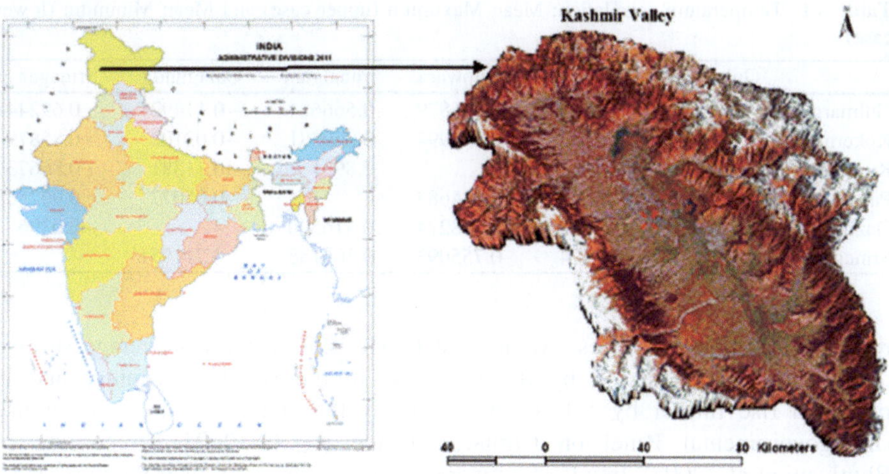

Fig. 8.1 Location map of the study area

Its floor stands 1,600 m above sea level in the Jhelum flood plain. It covers an area of about 4,865 km^2. The valley is accessible from the Punjab plain through two famous passes; the Pir Panjal pass (3,494 m) and Banihal pass (2,832 m) (Fig. 8.1).

8.3 Database and Methodology

The following hydro-meteorological data are used in the analysis. Temperature data (1975–2009) for 6 Meteorological Observatory stations were provided by the Indian Meteorological Department. The monthly discharge (1980–2009) of the Jhelum basin was provided by the Department of Flood Control (P & D Division) Srinagar, respectively. The statistical techniques used in this study to analyze the spatial variations and temporal trends of the hydro-climatic data series are (1) Karl Pearson's coefficient of correlation and (2) A simple linear regression method, parametric *t*-test method, to test the long-term linear trend. (3) To examine the relationship between temperature, precipitation and discharge, Kendall correlation analyses were performed.

8.4 Spatial Temperature Correlation of Kashmir Valley

One of the most prominent aspects of Kashmir's climate is its variability. This variability ranges over time and space from localized thunderstorms to larger scale features of fronts and storms. Mountainous regions are said to require a much greater density of climate stations than neighboring flatlands to achieve the same

Table 8.1 Temperature correlation: Mean Maximum (upper case) and Mean Minimum (lower case)

	Gulmarg	Kokernag	Kupwara	Pahalgam	Qazigund	Srinagar
Gulmarg	–	0.551127	0.546879	0.566684	−0.11492	0.692448
Kokernag	0.49494	–	0.895994	0.869381	−0.03102	0.858794
Kupwara	0.555118	0.676871	–	0.8951	−0.11381	0.866737
Pahalgam	0.392225	0.658508	0.634683	–	−0.05077	0.902568
Qazigund	0.543771	0.566388	0.578274	0.416596	–	−0.0396
Srinagar	0.445844	0.61174	0.755093	0.506158	0.711951	–

reliability of areal estimates (Archer and Fowler 2004). Earth's average surface temperature has increased by 0.6 ± 0.2 °C in the twentieth century and is projected rise further by 1.4 °C to 5.8 °C by the end of twenty-first century (Intergovernmental Panel on Climate Change (IPCC) 2001). According to Bhutiyani et al. (2007), trends of temperature are not uniform over the north west Himalayan region and annual temperature has risen by 1.6 °C in the last century.

Correlation matrixes between various stations of the valley are shown in Table 8.1. The climate of Kashmir valley has its own specialties ranging from small-scale phenomena of thunderstorms to more devastating floods. The stations are mostly positively correlated for mean maximum and mean minimum across the valley. Significant correlations are found between all stations of Kashmir valley except Qazigund for mean maximum temperature. Qazigund has shown negative correlation with all the other stations of the valley though not very significant. A significant high correlation was found for Kupwara, Pahalgam and Srinagar for mean maximum temperatures.

The correlation of the stations for mean minimum is also showing high degree of positive correlation. Most of the stations are strongly correlated with each other. The highest correlation is witnessed between Srinagar with Kupwara and Srinagar with Qazigund. There is no negative correlation found between any station of the valley and all the stations are significantly correlated.

8.5 Trends in Temporal Temperature Change

Trends for mean, mean maximum and mean minimum temperatures were investigated for the stations of Gulmarg, Kokernag, Kupwara, Pahalgam, Qazigund and Srinagar for the period 1975–2009. The records were analyzed by fitting a linear least squares trend line to the annual deviation. Student's t test was employed to assess the significance of trend. Trends for mean maximum and mean minimum temperatures we for the six stations of the valley are given in Table 8.2. Increases (+) and decreases (–) are shown as change in °C per decade. The trend is statistically significant ($p < 0.05$) at Gulmarg and Srinagar, with warming rates of +0.41 and +0.28 °C per decade, respectively (Table 8.2). Most of the stations are

Table 8.2 Trends for (A) mean (B) mean maximum and (C) mean minimum from 1975 to 2008, showing change in °C per decade

Station	Annual	Winter (DJF)	Spring (MAM)	Summer (JJA)	Autumn (SON)
(A)					
Gulmarg	**+0.41**	***+0.23***	−0.12	**+0.41**	+0.19
Kokernag	+0.11	+0.13	**+0.21**	***−0.47***	**+0.49**
Kupwara	−0.17	−0.09	−0.13	**+0.39**	***−0. 25***
Pahalgam	***+0.21***	**+0.29**	**+0.50**	***−0.25***	−0.21
Qazigund	−0.11	+0.12	−0.07	+0.20	+0.13
Srinagar	**+0.28**	**+0.35**	+0.14	**+0.39**	+0.06
(B)					
Gulmarg	+0.12	**−0.28**	−0.10	+0.01	+0.11
Kokernag	−0.02	**−0.29**	−0.05	−0.02	−0.24
Kupwara	+0.18	+0.13	−0.11	**+0.23**	**+0.37**
Pahalgam	**−0.34**	−0.07	***−0.26***	+0.13	+0.31
Qazigund	**+0.25**	+0.12	**+0.43**	**−0.46**	+0.21
Srinagar	**+0.36**	***+0.27***	+0.21	**−0.63**	**+0.59**
(C)					
Gulmarg	**−0.49**	***+0.31***	−0.26	−0.13	−0.08
Kokernag	**−0.38**	**+0.42**	−0.12	***−0.23***	−0.13
Kupwara	−0.03	+0.12	−0.17	−0.22	+0.26
Pahalgam	**−0.31**	***−0.16***	**−0.69**	**−0.42**	**−0.58**
Qazigund	+0.07	+0.07	**+0.38**	−0.12	+0.18
Srinagar	+0.17	+0.14	+0.21	−0.09	**+0.48**

Bold $p < 0.05$. Bold italic $p < 0.10$

observing significant warming of annual mean temperatures. Winter temperatures are observed to show a warming trend at stations of Srinagar (+0.35), Pahalgam (+0.29) and Gulmarg (+0.23) per decade respectively. Summer season witnessed a positive summer temperature trend with rates of +0.41, +0.39 and +0.39 decade^{-1} for the stations of Gulmarg, Kupwara and Srinagar respectively. Small but significant cooling trends are bserved for the stations of Kokernag (−0.47) and Pahalgam (−0.25) per decade. Similarly, spring mean temperature is showing warming trend at three out of six stations. But it is only significant at Pahalgam and Kokernag with +0.50 and +0.21 warming per decade respectively. The other stations do not show any significant trend for the said season.

There has been an increase in annual maximum temperature at all stations except Kokernag and Pahalgam (Table 8.2). Significant increases in maximum temperatures occur mainly in autumn season whereas decrease in maximum temperatures occurs in spring season. There has been significant decrease in maximum temperature in summer months in Qazigund (−0.46) and Srinagar (−0.63) whereas Kupwara has observed an increase of +0.23 per decade. The most significant increase for the autumn season is observed by Kupwara (+0.37) and Srinagar (+0.59). Parthasarathy and Dhar found a positive trend over Central India and the adjoining parts and a decreasing trend over some eastern parts of India for the period 1901–1960. Brown et al. (2008) found that extreme daily maximum and minimum temperatures have warmed for most regions of the world since 1950 and that the positive trends is significantly greater than that which can be attributed to natural variability.

Most of the stations are observing significant cooling of minimum temperatures. Gulmarg, Kokernag and Pahalgam are witnessing a cooling rates of −0.49, −0.38 and −0.31 per decade respectively (Table 8.2). Surprisingly there is decrease in summer minimum temperature, most significant in Kokernag (−0.23) and Pahalgam (−0.42). The station of Pahalgam witnessed a decreasing minimum summer temperature with rate of −0.58 per decade. Meanwhile, Srinagar has observed an increase of +0.48 per decade. Significant winter temperature increase is observed in Gulmarg (+0.31) and Kokernag (+0.42) stations.

8.6 Trends in Precipitation Change in the Upper Jhelum Basin

Though each and every part of the world is more or less susceptible to natural calamities, the Himalaya due to its complex geological structures, dynamic geomorphology, and seasonality in hydrometeorological conditions experience natural disasters very frequently, especially water induced hazards (Rawat et al. 2011). The overall association of the stations is analyzed by fitting a linear least square trend line to the annual deviation from the mean. Seasonal, summer and winter trends are shown in Fig. 8.2.

The stations of Srinagar, Kokernag and Pahalgam are witnessing a significant downward annual trend. There is a sharp decreasing trend for the station of Pahalgam. The rainfall total of these stations is small but most variable. The precipitation in the winter season is received by virtue of the cyclones which develop in the Mediterranean Sea. The only first class city of the area (Srinagar) has witnessed a significant downward trend. According to Wani and Khairkar (2012), the seasonal mean minimum temperature series of Srinagar indicate that the winter season and spring season show warming trend especially for winter season. The total built up of Srinagar city has increased by 159.61 % whereas the forest cover has drastically decreased by 55 % during the last 40 years in Srinagar city (Wani and Khairkar 2011). The study shows a slow but continous decrease in the amount of precipitation in the area.

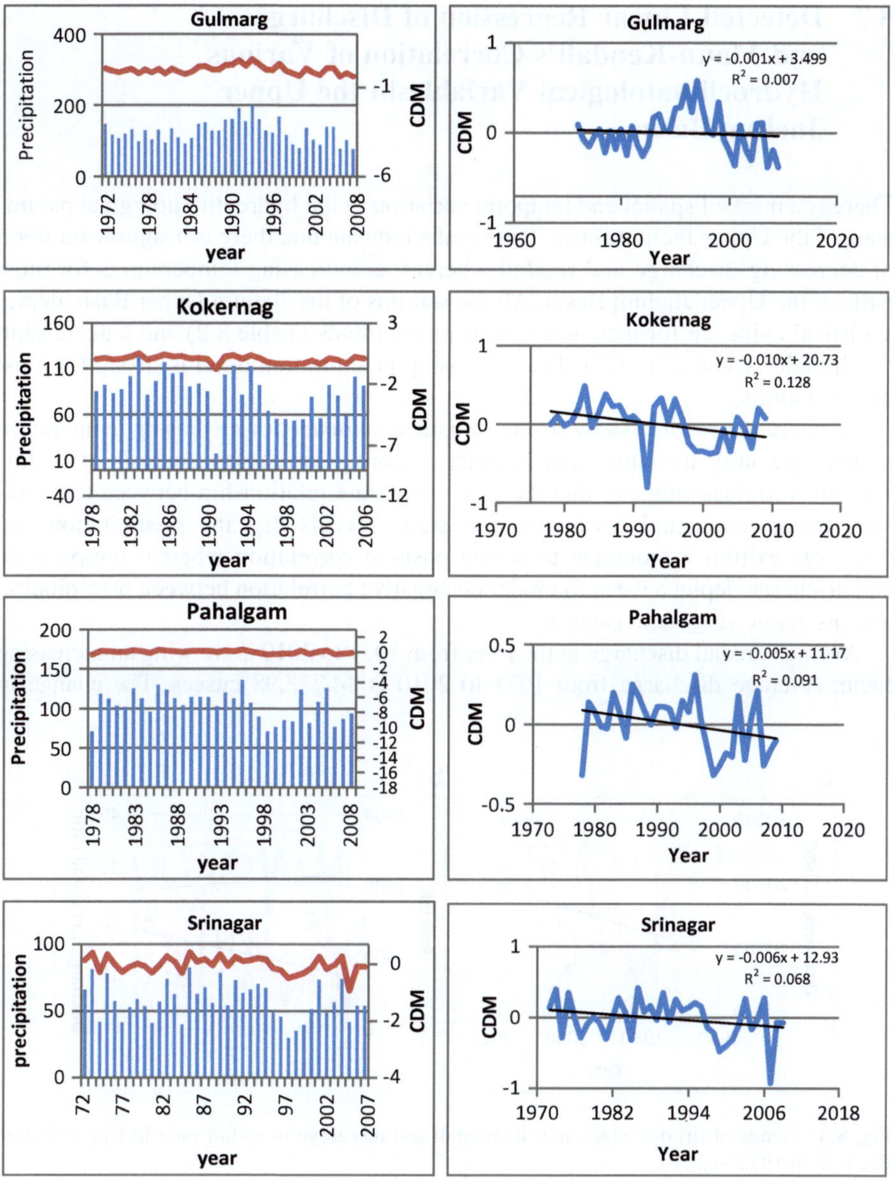

Fig. 8.2 Trends of precipitation from 1972 to 2009 in Upper Jhelum Basin catchment. *Right hand* shows the deviation from the mean annual precipitation, with a least squares fitted trend and *left hand* shows annual precipitation annual precipitation with cumulative departure from mean plotted as a *line*

8.7 Detected Linear Regression of Discharge and Mann-Kendall's Correlation of Various Hydroclimatological Variablesin the Upper Jhelum Basin

There is a marked spatial and temporal variation in the hydroclimatological parameters of the Upper Jhelum Basin. Our results indicate that there is a significant trend of decreasing discharge and rainfall whereas as increasing temperatures for most parts of the Upper Jhelum Basin. All the stations of the Jhelum Upper Basin depict statistical evidence for increasing temperature trends (Table 8.2) and a decreasing precipitation trend (Fig. 8.2). The decreasing precipitation trend is strong for most of the stations.

Temperature records show strong statistical evidence for increasing temperature trends especially in winter and summer seasons. Correlation analysis of hydro-climatic variables suggests that there is a moderate relationship between temperature, precipitation and discharge. The study reveals that the precipitation and discharge exhibit a moderate to strong positive correlation whereas temperature and discharge depict a weak to moderate negative correlation between precipitation and discharge (Fig. 8.3, Table 8.3).

Average annual discharge in the river from 1971 to 2010 is showing an increasing trend. Average discharge from 1971 to 2010 is 14,733.98 cusecs. The change in

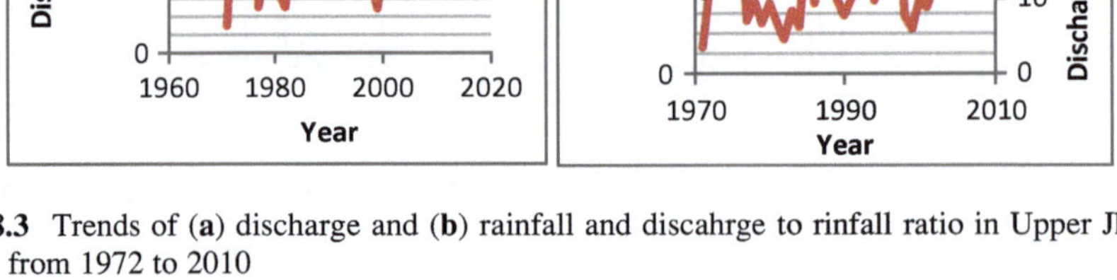

Fig. 8.3 Trends of (**a**) discharge and (**b**) rainfall and discahrge to rinfall ratio in Upper Jhelum Basin from 1972 to 2010

Table 8.3 Kendalls correlation between discharge, temperature and preciptation

	Discharge/temperature					Discharge/precipitation				
	An.	Sp.	Sum	Aut	Win	An.	Sp.	Sum	Aut	Win
Kendall's value	−0.13	0.12	−0.33	−0.28	0.28	**0.37**	0.15	0.22	−0.17	**0.52**
p	0.02	0.11	0.16	0.08	0.12	0.06	0.07	0.18	0.00	0.01

An. annual, *Sp.* Spring, *Sum.* Summer, *Aut.* Autumn, *Win.* Winter average annual discharge from 1971 to 2010 has been negative for 26 years with respect to the average discharge i.e. for 15 years average annual discharge has remained below the average discharge. The average annual discharge along with the discharge to rainfall ratio is manifesting an increasing trend. Since the annual precipitation trends is slightly decreasing, temperature trends appear to be the factor responsible for discharge trends to increase which clearly supports the previous studies of glacier retreat and global warming. The mean minimum temperature of Srinagar has increased by almost 3 °C during the last 40 years (wani and Khairkar 2011).

8.8 Conclusion

In this study, temporal trends of precipitation and temperature over Kashmir valley especially in the Upper Jhelum Catchment area were examined. The temporal trends of drainage and the impact of climate change on discharge in the Upper Jhelum catchment area were also analyzed. The stations across the valley are mostly positively correlated for mean maximum and mean minimum. A significant high correlation was found for Kupwara, Pahalgam and Srinagar for mean maximum temperatures. Temperature data show significant trend for temperature in some parts of the Kashmir valley at the $\alpha = 0.05$ significance level. The spatial distribution patterns of temperature trends vary seasonally. The study reveals that there is a significant increase in mean summer temperatures whereas decreases in summer mean minimum temperatures. There is a significant increasing trend observed by mean annual temperatures whereas the annual mean minimum is witnessing a significant decreasing trend. The study also reveals a decreasing trend for precipitation in most parts of the Jhelum basin. The study reveals that the precipitation and discharge exhibit a moderate to strong positive correlation whereas temperature and discharge depict a weak to moderate negative correlation between precipitation and discharge.

References

Archer DR, Fowler HJ (2004) Spatial and temporal variations in precipitation in the Upper Indus Basin, global teleconnections and hydrological implications. Hydrol Earth Syst Sci 8(1):47–61

Bhutiyani MR, Kale VS, Pawar NJ (2007) Long-term trends in maximum, minimum and mean annual air temperatures across the Northwestern Himalaya during the twentieth century. Clim Change 85:159–177

Brown SJ, Caesar J, Ferro CAT (2008) Global changes in extreme daily temperature since 1950. J Geophys Res 113, D05115. doi:10.1029/2006JD008091

Intergovernmental Panel on Climate Change (IPCC) (2001) Climate change 2001: the scientific basis. In: Houghton JT, Ding Y, Griggs DJ, Noguer M, van der Linden PJ, Dai X, Maskell K, Johnson CA (eds) Contributions of Working Group I to the third assessment report of the intergovernmental panel on climate change. Cambridge Univ. Press, New York

IPCC (1995) Climate Change 1995: the science of climate change. In: Houghton JT et al. (eds) Cambridge University Press, Cambridge

Jones PD, Moberg A (2003) A hemispheric and large-scale surface air temperature variations: an extensive revision and an update to 2001. J Climate 16:206–223

Pant GB, Borgaonkar HP, Rupa Kumar K (1998) Climatic signals from tree rings: a dendroclimatic investigation of Himalayan spruce (*Picea smithiana*). Himalayan Geol 19:65–73

Rawat PK, Pant CC, Tiwari PC, Pant PD, Sharma AK (2011) Spatial variability assessment of river-line floods and flash floods in Himalaya: A case study using GIS. Int J Disast Prev Manage 21(2):135–159

Sahsamanoglou HS, Makrogiannis TJ (1992) Temperature trends over the Mediterranean region, 1950–1988. Theor Appl Climatol 45:183–192

Singh RB, Roy SS (2002) "Climate variability and hydrological extremes in a Himalayan Catchment" ERB and Northern European FRIEND Project 5 Conference, Demänovská dolina, Slovakia

Tsonis AA, Swanson K, Kravisov S (2007) A new dynamical mechanism for major climate shifts. Geophys Res Lett 34:L13705. doi:10:1029/GL030288

Wani RA, Khairkar VP (2011) Quantifying land use and land cover change using geographic information system: A case study of Srinagar city, Jammu and Kashmir, India. Int J Geomatics Geosci 2(1):110–120

Wani RA, Khairkar VP (2012) "Climatic Variability over Srinagar city: identifying trends" Researchers world. J Arts Sci Commerce III(1):142–160

Xu J (2009) The trend of land use and land cover change and its impacts on biodiversity in the Himalayas. In: International Mountain Biodiversity Conference, www.icimod.org/resource.php?id=103

Chapter 9
Carbon Sequestration in Soils of Different Land Use System in Sirsi Taluka of Uttara Kannada District Through Geo-Informatics Approach

A.G. Koppad and Pavan Tikhile

Abstract The study was conducted in Sirsi taluka (14°28′0″ N to 14°51′30″ N Latitude and 74°34′0″ E to 75°03′30″ E Longitude) of Uttar Kannada district of Karnataka in India to assess the influence of forests on soil organic carbon. IRS P6 LISS-III imageries for the study area was procured from NRSC, Hyderabad and different land use systems in Sirsi taluka were identified with the ground truth data processed in ERDAS software. The land use classes viz., dense forest, sparse forest, plantation, Agriculture and open land were identified. The total area in each class was assessed through supervise classification. The soil samples at one meter depth were drawn at grid point in flat land and along the profile in sloppy land in different land use system. The SOC was estimated using Walkley and Black rapid titration method. The pattern of Organic carbon distribution in Sirsi taluka is determined using Inverse Distance Weight (IDW) technique in Arc GIS software. The total area in five land use classes is 1,29,964 ha with SOC of 32.38 million tonnes. Among different classes dense forest covers highest area (63,763 ha) and highest SOC pool (21.38 million tonnes). Among the different land use classes, higher SOC was sequestered in horticulture plantations (361.05 t/ha) followed by Dense forest (335.25 t/ha). The SOC in sparse forest and barren land is 239.39 t/ha and 168.74 t/ha respectively. The lowest SOC was recorded in Agriculture land (76.50 t/ha). The CO_2 mitigation potential of horticulture plantation is 4.72 times higher followed by dense forest (4.38 times), sparse forest (3.13 times) and open land (2.21 times) as compared to agriculture land. The study indicated that horticulture plantation and dense forest are the sinks of carbon.

Keywords Forest • IDW • Imageries • LULC • Soil organic carbon

---

A.G. Koppad (✉) • P. Tikhile
Department of Agricultural Engineering, Agricultural College, Bijapur 586101, India
e-mail: koppadag2007@rediffmail.com

M. Singh et al. (eds.), *Climate Change and Biodiversity: Proceedings of IGU Rohtak Conference, Vol. 1*, Advances in Geographical and Environmental Sciences,
DOI 10.1007/978-4-431-54838-6_9,

9.1 Introduction

Global warming is become a problem for the society due to carbon emission in the wake of modernization and urbanization. Global surface temperature has increased by 0.8 °C since the late nineteenth century and 11 out of the 12 warmest years on record have occurred since 1995 (IPCC Climate Change 2007). The soil plays a vital role in global carbon cycle, Carbon (C) stored in soil is 2.5–3.0 times higher than that of plants (Post et al. 1990) and two to three times more than the atmospheric CO_2 (Davidson et al. 2000). Soil organic carbon (SOC) stock acts as a major part of the terrestrial carbon reservoir (Grace 2004) with a storage of about 1,500 Pg to 2,000 Pg C (1 Pg $= 10^{15}$ g, or 1 Pg = 1 billion tonnes) in the top 100 cm depth layer in the world soils (Batjes 1996; Yu et al. 2012) so small changes in the terrestrial SOC pool can have a large effect on C cycle (Schlesinger and Andrews 2000) and thus constitute a powerful positive feedback to the climate system (Jandl et al. 2007; Koppad and Tikhile 2013). Land use land cover changes directly affect the carbon sequestration rate in soil (Lal 2004a) and other factors including climate, vegetation type, topography and anthropogenic activities (Six and Jastrow 2002; Baker 2007). The forest soils are one of the major carbon sink in earth having higher content of organic carbon (Dey 2005). The soil bulk density plays a very important role in the assessment of SOC contents (Han et al. 2010). There is a major potential for increasing SOC through restoration of degraded soils and widespread adoption of soil conservation practices (Lal and Bruce 1999, Lal et al. 1997). Carbon sequestration has the potential to offset fossil fuel emissions by 0.4–1.2 gigatons of carbon per year, or 5–15 % of the global fossil-fuel emissions (Lal 2004b).

Forest and Forest soil play an important role in the global carbon balance (Kuldeep and Upasana 2011). About 2 Gt C/year is being removed from the atmosphere and stored in vegetation and soil (Bolin et al. 2000; Lenton 2000). The sequestered carbon finally acts as a sink in the forest land (Ramchandran et al. 2007). Accurate estimation of soil organic carbon pool helps for climate change study and prediction of variation in climate change and its effect on local and global climate as well as biodiversity. The remote sensing technologies are being used for preparation of land use land cover (LULC) map and area estimation under different land use practices. Geographical information system (GIS) technology is useful for spatial analysis and preparation of SOC distribution map. The present study is an attempt to estimate the carbon sequestration stock in forest soils of various land use in Sirsi taluk of Uttar Kannada district of Karnataka.

9.1.1 Carbon Sequestration

Carbon is found in all living organisms, it is the major building block for life in Earth. Carbon exists in many forms, predominately as plant biomass, soil organic carbon and as carbon dioxide in the atmosphere and dissolved in seawater.

Soil carbon sequestration is the process of transferring carbon dioxide from the atmosphere into the soil (Roger and Brent 2012, Watson et al. 2000) through crop residues and other organic solids which is not immediately re-emitted, which helps off-set emissions from fossil fuel combustion and other carbon-emitting activities. Through the process of photosynthesis plants assimilate carbon and return to the soil as litter and stored as soil organic matter (Negi and Gupta 2010). Soil organic carbon includes plant, animal and microbial residues in all stages of decomposition (Post and Kwon 2000). Tree growth serve as an important mean to capture and store atmospheric CO_2 in vegetation, soil and biomass product (Makundi and Sathaye 2004). Plants produce organic compounds by using sunlight energy and combining carbon dioxide from the atmosphere. Soil organic matter is created by the cycling of these organic compounds in plants, animals and microorganisms into the soil. Carbon is a key ingredient in soil organic matter (57 % by weight). As the soil is major sink of carbon, the amount of carbon storage in soil mainly depends on contribution of organic matter from the forest, its decomposition, its retention in the soil which in turn depends on topographic and environment factors mainly, land slope, drainage system, total rainfall and rainfall intensity. Carbon sequestration in biomass and in soil is a key strategy to reduce atmospheric CO_2 (Sundaram et al. 2012).

9.1.2 National Status

The total geographical area of India is 328.7 million hectares (Mha) or about 2.5 % of the total land area of the world (Lal 2004c). India ranks 10th in the list of most forested nations in the world with 76.87 Mha of forest and tree cover acquiring 23.4 % total geographical area of the country (Kishwan et al. 2009). The first estimate of the organic carbon pool in the Indian soil done in the year 1984 and it was 24.3 pg (1 Pg $= 10^{15}$ g) based on 48 soil sample (Gupta and Rao 1984). India first 30 cm of soil holds 9 Pg (Bhattacharya et al. 2000). In India, total and forest SOC stocks are in the range of 23.4–47.5 Pg C (Dadhwal and Nayak 1993, Gupta et al. 1993) and 5.4–6.7 Pg C (Ravindranath et al. 1997.) respectively. SOC based on different forest types are estimated using bulk density calculated indirectly (Chhabra et al. 2002). SOC ranges between 140.76 t/h to 58.45 t/h recorded in different forest types in the state of Uttarakhand (Negi et al. 2013). According to Jha et al., in 27 states of India total SOC pool is 9,815.95 million tonnes out of 539.32 million tonnes (Forest area 3.1818 million ha) in Karnataka (Jha et al. 2003). The Himalayan zones, with dense forest vegetation, cover nearly 19 % of India and contain 33 % of SOC reserves of the country (Bhattacharyya et al. 2008). In India, attempts were made to assess carbon sequestration studies at macro level (Ravindranath et al. 1997) mostly with the available data. No attempt has been made so far to assess the soil carbon sequestration at micro-level. Such kind of micro level study is essential for sustainable forest management, especially in a country like India, where heavy degradation had been caused by anthropogenic activities and different forest management prescriptions of the past warranted in different periods of time to meet the local and national needs.

9.2 Material and Method

The study was conducted in Sirsi taluka of Uttar Kannada district of Karnataka, India. The study region lies between 14°28′0″ N to 14°51′30″ N Latitude and 74°34′0″ E to 75°03′30″ E Longitude covering a surface area about 1,31,830 ha and with average elevation of 590 m from Mean Sea level. The study area is shown in Fig. 9.1.

The topo-sheets covering the study area are 48 J/9, 48 J/10, 48 J/11, 48 J/12, 48 J/13, 48 J/14, 8 J/15, 48 N/2, and 48 N/3 with scale 1:50,000 were procured from SOI Bangalore. IRS LISS-3 satellite image path 097 row 063 dated 22 January 2010 with spatial resolutions of 23.5 × 23.5 m were procured from NRSC Hyderabad. The images were processed using ERDAS IMAGINE 2011 and classify the land use land cover classes and spatial analysis of SOC was done using IDW technique in Arc GIS 10.

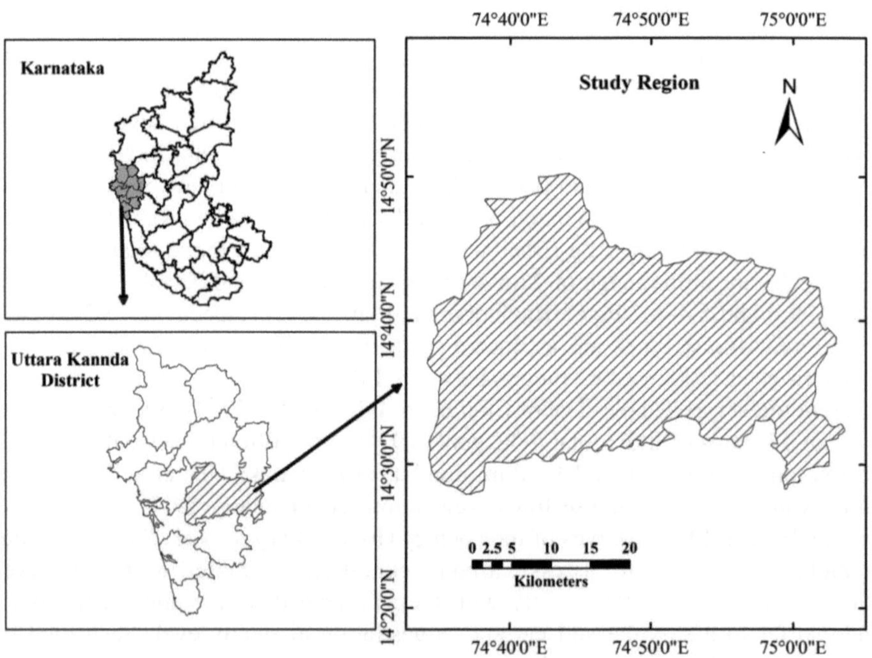

Fig. 9.1 Location map of study area

Soil samples were collected from different land use land cover classes areas viz., Dense forest, sparse forest, plantation, agriculture and open land. While taking the samples, the land terrain was considered. In sloppy land the number of samples were taken along the slope from top to bottom and in flat land representative soil sample at different grids was collected. The soil sample spot latitude and longitude was recorded with GPS. The soil sample to the depth of 1 m was taken using soil screw auger. The core sampler was used to collect the soil core at different depth for estimating the soil bulk density. SOC was determined using Walkley and Black rapid titration method (Walkley and Black 1934). The % of SOC value obtained from the WB method was multiplied by standard correction factor of 1.32 (DeVos et al. 2007) to obtain the corrected SOC.

$$\text{SOC}\,\% = (\text{BTV-STV}) \times 0.5\text{N FASx}0.003 \times 100/\text{wt. of soil}$$

$$\text{Total SOC (tones)} = \text{SOC}\,\%/100 * \text{BD}\left(\text{t/m}^3\right) * \text{area}\left(\text{m}^2\right) * \text{depth of soil (m)}$$

BTV = Blank titrated value, STV = Sample titrated value, FAS = Ferrous Ammonium Sulphate

BD = Bulk density.

SOC (million tonnes) = Area (ha) × depth (m) × Bulk density (t/m^3) × SOC (%) × 10^{-4}

The ground truth data was collected using GPS and the same data was used as ground truth data in image classification for preparing land use land cover thematic map. Soil samples collected from different land use systems with GPS locations were analyzed for SOC. The SOC map was prepared using IDW interpolation method.

9.3 Result and Discussion

The land use land cover classification indicated that about 48.37 % (63,763 ha) of the areas is occupied by dense forest followed by Spares forest (34,936.60 ha), Agriculture (21,715.30 ha), Plantation (4,437.54 ha), Open land (5,112.39 ha), Settlement (1,279.95 ha) and Water body (586.02 ha) as shown in Fig. 9.2.

The SOC distribution map in Sirsi taluka prepared using IDW technique is shown in Fig. 9.3. The maximum SOC is found in plantation which is 2.27 % followed by dense forest (2.25 %) and very less was found in agriculture soil (0.75 %). SOC distribution map indicated that the north, north-east, west and south-west region of study area having higher percentage of soil organic carbon mainly due to dense forest and horticulture plantation. The distribution of SOC in south and east area is very less due to agriculture and open land.

Soil organic carbon pool under different land use classes is presented in Table 9.1. The results indicated that the horticulture plantation is having higher SOC pool (361.05 t ha^{-1}) followed by dense forest (335.25 t ha^{-1}), Sparse forest

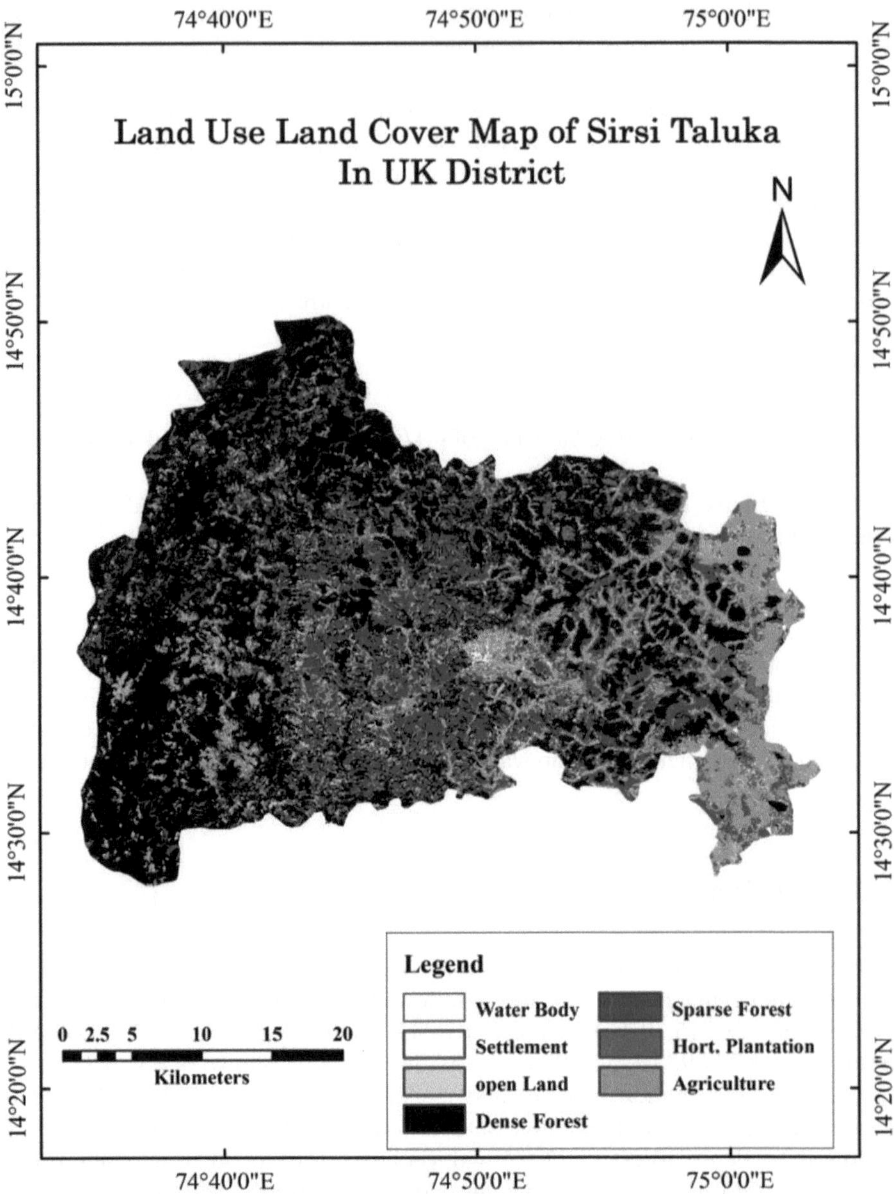

Fig. 9.2 LULC map of Sirsi taluka

(239.39 t ha^{-1}), barren land (168.74 t ha^{-1}) and least was found in agriculture land (76.50 t ha^{-1}).

The Carbon sequestration and mitigation potential was worked out for all land use land cover classes as compared to agriculture land. The result showed that the

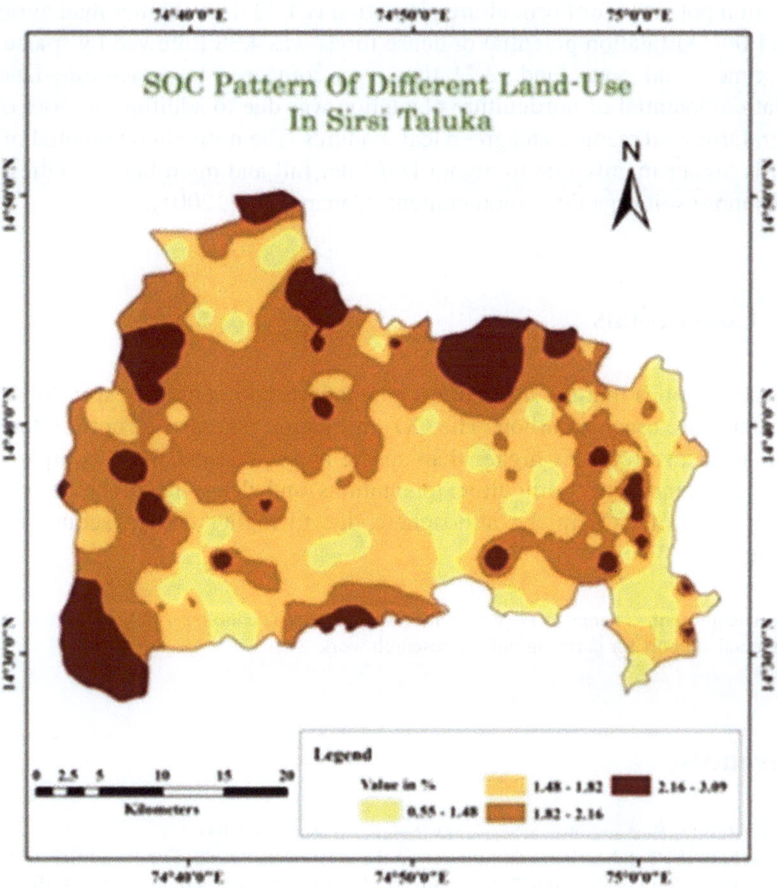

Fig. 9.3 SOC pattern in Sirsi taluka

Table 9.1 Total SOC (million tonnes) from different LULC in Sirsi taluka of UK district

Sr. no	Land use class name	Area (ha)	Area (%)	SOC (%)	B.D (t/m^3)	Total SOC (million tonnes)	SOC (t/ha)	Mit.pot LU wise
1	Horticulture plantation	4,437.54	3.37	2.27	1.15	1.1584	361.05	4.72
2	Dense forest	63,763.00	48.37	2.25	1.49	21.3765	335.25	4.38
3	Sparse forest	34,936.60	26.50	1.71	1.40	8.3638	239.39	3.13
4	Open land	5,112.39	3.88	1.35	1.25	0.8627	168.74	2.21
5	Agriculture	21,715.30	16.47	0.75	1.02	1.6612	76.50	1.00
6	Settlement	1,279.95	0.97	–	–	–		–
7	Water body	586.02	0.44	–	–	–		–
	Total	131,830.80	100			32.3831		

mitigation potential for horticulture plantation is 4.72 times higher than agriculture land (1.00). Mitigation potential of dense forest was 4.38 followed by sparse forest (3.13 times) and open land (2.21 times) as compared to agriculture land. The mitigation potential of horticulture plantation was due to addition of more organic matter, farm yard manure and green leaf manures. The mitigation potential of dense forest is higher mainly due to higher leaf litter fall and microbial activities which creates more soil organic carbon content (Conanat et al. 2001).

9.4 Conclusions

The study implies that the horticulture plantations and dense forest in Sirsi taluka sequester more organic carbon. The CO_2 mitigation potential of horticulture plantation and dense forest is higher than sparse and open land when compared with agricultural land. The horticulture plantations and dense forest helps to reduce concentration of CO_2 in the atmosphere due to higher carbon sequestration in the soil.

Acknowledgement Thanks to Department of Science and Technology, New Delhi for providing the financial support for carrying out the research work.

References

Baker DF (2007) Reassessing carbon sinks. Science 316:1708–1709

Batjes NH (1996) Total carbon and nitrogen in the soils of the world. Eur J Soil Sci 47:151–163

Bhattacharya T, Pal DK, Mondal C, Velayutham M (2000) Organic carbon stock in Indian soil and their geographical distribution. Curr Sci 79(5):655–660

Bhattacharyya T, Pal DK, Chandran P et al (2008) Soil carbon storage capacity as a tool to prioritize areas for carbon sequestration. Curr Sci 95:482–494

Bolin B, Sukumar R, Ciais P et al (2000) Global perspective. In: Watson RT, Noble IR, Bolin B, Ravindranath NH, Verardo DJ, Dokken DJ (eds) Land use, land-use change and forestry. A Special Report for the IPCC. Cambridge University Press, Cambridge, pp 24–51

Chhabra A, Palria S, Dadhwal VK (2002) Growing stock based forest biomass in Indian forests. Biomass Bioenergy 22:187–194

Conanat RT, Paustian K, Elliott ET (2001) Grassland management and conversion into grassland: effect on soil carbon. Ecol Appl 11:343–355

Dadhwal VK, Nayak SR (1993) A preliminary estimate of bio-geochemical cycle of carbon for India. Sci Cult 59:9–13

Davidson EA, Trumboe SE, Amudson R (2000) Soil warming and organic carbon content. Nature 408:789–790

DeVos B, Lettens S, Muys B, Deckers JA (2007) Walkley-Black analysis of forest soil organic carbon: recovery, limitation and uncertainty. Soil Use Manage 23:221–229

Dey SK (2005) A Preliminary estimation of carbon stock sequestrated through rubber (*Hevea brasiliensis*) plantation in North Eastern regional of India. Indian Forester 131(11):1429–1435

Grace J (2004) Understanding and managing the global carbon cycle. J Ecol 92:189–202. doi:10.1111/j.0022-0477.2004.00874.x

Gupta AK, Nageswara Rao PP, Ganesha Raj K, Dutt CBS, Chandrasekhar MG (1993) An integrated approach for resources development using remote sensing: a case study specific to hill area. J Indian Soc Remote Sens 21:99–108
Gupta RK, Rao DLN (1984) Potential of wastelands for sequestering carbon by reforestation. Curr Sci 66:378–380
Han F, Wei H, Zheng J et al (2010) Estimating soil organic carbon storage and distribution in a catchment of Loess Plateau, China. Geoderma 154:261–266
IPCC Climate Change (2007) Climate change impacts, adaptation and vulnerability. In: Working Group II 2007 Geneva. IPCC, Switzerland
Jandl R, Vesterdal L, Olsson M et al (2007) Carbon sequestration and forest management. CAB Reviews 2(17). doi:10.1079/PAVSNNR20072017
Jha MN, Gupta MK, Saxena A, Kumar R (2003) Soil organic carbon store in different forests of India. Indian forester 129(6):714–724
Kishwan J, Pandey R, Dadhwal VK (2009) India's forest and tree cover: contribution as a carbon sink. Technical Paper. Indian Council of Forestry Research and Education, Deharadun
Kuldeep P, Upasana P (2011) Forest carbon management using satellite remote sensing techniques. A case study of Sagar district. E Int Sci Res J 3(4):335–348
Koppad AG, Tikhile P (2013) Influence of land use land cover classes on carbon sequestration in soils of Sirsi and Siddapur taluka of Uttara Kannada District, India. Int J Curr Res 5(4): 1012–1015
Lal R (2004a) Soil carbon sequestration to mitigate climate change. Geoderma 123:1–22
Lal R (2004b) Soil carbon sequestration impacts on global climate change and food security. Science 304:1623–1627
Lal R (2004c) Soil carbon sequestration in India. Clim Change 65:277–296
Lal R, Bruce (1999) Soil management and restoration for C sequestration to mitigate the accelerated greenhouse effect. Prog Environ Sci 1:307–326
Lal R, Kimble J, Follett R (1997) Land use and soil C pools in terrestrial ecosystems. In: Lal R (ed) Management of carbon sequestration in soil. CRC, Florida, pp 1–10
Lenton TM (2000) Land and ocean carbon cycle feedback effects on global warming in a simple earth system model. Tellus 52B:1159–1188
Makundi WR, Sathaye JA (2004) GHG mitigation potential and coast in tropical forestry relative role for agro-forestry. Environ Dev Sustain 6:235–260
Negi SS, Gupta MK (2010) Soil organic carbon store under different land use systems in Giri catchment of Himachal Pradesh. Indian Forester 136(9):1147–1154
Negi SS, Gupta MK, Sharma SD (2013) Sequestered organic carbon pool in the forest soils of Uttarakhand state, India. Int J Sci Environ Tech 2(3):510–520
Post WM, Pengh TH, Emanuel W, King AW, Dale VH, Delnglis (1990) The global carbon cycle. Am Sci 78:310–326
Post WM, Kwon KC (2000) Soil carbon sequestration and land-use change: processes and potential. Glob Chang Biol 6:317–328
Ramchandran A, Jaykumar S, Haroon RM et al (2007) Carbon sequestration: estimation of carbon stock in natural forest using geospatial technology in the Eastern Ghat of Tamil Nadu, India. Current Sci 92(3):323–331
Ravindranath NH, Somashekhar BS, Gadgil M (1997) Carbon flows in Indian forests. Climatic Change Biol 35:297–320
Roger S, Brent S (2012) Carbon Sequestration in Forests and Soils. Annu Rev Resource Econ 4: 127–144 doi:10.1146/annurev-resource-083110-115941
Schlesinger WH, Andrews JA (2000) Soil respiration and the global carbon cycle. Biogeochemistry 48:7–20
Six J, Jastrow JD (2002) Organic matter turnover. In: Lal R (ed) Encyclopedia of soil science. Marcel Dekker, New York, pp 936–942
Sundaram SM, Sivakumar T, Sankaran VM et al (2012) Distribution of soil organic carbon (SOC) in agricultural lands by farming forage crops for mitigating climate change. Int J Environ Biol 2(3):165–168

Walkley A, Black IA (1934) An examination of Degtjareff method for determining organic carbon in soil: effect of variation in digestion condition of inorganic soil constitution. Soil Sci 63:251–263

Watson RT, Noble IR, Bolin B et al (2000) Land use, land change and forestry. Special: report of the intergovernmental panel on climate change. Cambridge University Press, Cambridge, pp 111–161

Yu J, Wang Y, Li Y et al (2012) Soil organic carbon storage changes in coastal wetlands of the modern Yellow River Delta from 2000 to 2009. Biogeosciences 9:2325–331

Part II
Biodiversity

Chapter 10
Land Cover Change and Rhino Habitat Mapping of Kaziranga National Park, Assam

Ankita Medhi and Ashis Kumar Saha

Abstract Kaziranga National Park in Assam is a habitat for the highest population of one-horned rhino in the world. Conservation management of the park has become a serious concern to maintain the wildlife in the park. The present study investigates land cover change within Kaziranga National Park during the last two decades (1990–2009) using remote sensing-GIS techniques and analyses habitat suitability for rhino to understand possible effect of land cover change on the rhino habitat. The change detection analysis has shown considerable reduction of grasslands areas and small water bodies. Furthermore, Habitat Suitability Model for rhino has been developed based on semi-quantitative Analytical Hierarchy Process. The result shows decline in the suitable habitats for rhino during this period. Assessment of rhino habitat change indicates that any change in the land cover trigger substantial change in the suitable habitats for rhino. Moreover, increase in rhino population as reported by census and reducing suitable habitat further limit the carrying capacity of the national park. The results from the present study may be used as baseline for future rhino habitat monitoring.

Keywords Habitat suitability model • Kaziranga National Park • Landcover change • One-horned Rhinoceros

10.1 Introduction

Kaziranga National Park (KNP) in Assam, comprising of world's two third of one-horned rhino population is one of the important one-horned Rhinoceros (*Rhinoceros unicornis*) habitats in the world. The park is located on the flood plains

---

A. Medhi • A.K. Saha (✉)
Department of Geography, Delhi School of Economics, University of Delhi, Delhi, India
e-mail: ashisksaha@gmail.com

M. Singh et al. (eds.), *Climate Change and Biodiversity: Proceedings of IGU Rohtak Conference, Vol. 1*, Advances in Geographical and Environmental Sciences,
DOI 10.1007/978-4-431-54838-6_10, © Springer Japan 2014

of the Brahmaputra River. The Asiatic rhino species prefer to reside in the alluvial floodplain and vegetation of sub-tropical climate where water and green grasses are available year round. KNP is also well known for hundreds years of success in rhino conservation. However, conservation management of the park has become a serious concern for the last two decades. Though the rhino population is increasing at a positive rate in Kaziranga from 366 (in 1966) to 2,048 (in 2009), the major concern is the availability of suitable habitats in the park. The park may be reaching its carrying capacity for rhino. Moreover, the park area is severely affected by recurring monsoonal floods and rhino population is also threatened by illegal poaching and invasive species *Mimosa invisa*. Therefore, detailed monitoring to assess the spatial and temporal changes in the park is essential for proper management of the park.

National parks and protected areas face serious conservation management challenges due to changing land cover types and variability of landscape contexts within and adjacent to the park boundaries (Wang et al. 2009). Remote sensing provides a broad view of landscapes and can be consistent through time, making it an important tool for monitoring and managing protected areas and national parks (Kennedy et al. 2009; Gross et al. 2009; Wang et al. 2009). Understanding of the pattern of landuse/landcover change in national park would help park managers to evaluate the status and trends of the park to make strategies for effective conservation. Several studies on land cover mapping of National Parks have been carried out highlighting the applicability of remote sensing and GIS techniques (Kushwaha and Unni 1986; Saxena 1986; Kachhwaha 1993; Nagendra et al. 2004; Sarma et al. 2008; Bayarsaikhan et al. 2009).

Land cover mapping plays an important role in habitat suitability mapping as the habitat variables of any species depends on different land cover types. Habitat suitability mapping for rhino is required for the assessment of rhino habitat in Kaziranga National Park. The predictive model of habitat suitability for rhino is required to analyze the potential sites for rhino habitat. Several studies on habitat suitability have been done in India using remote sensing and GIS techniques. Habitat assessment of some ungulates in Kaziranga National Park (Parihar et al. 1986), habitat analysis for sambar in Corbett National Park(Pant et al. 1999), geospatial modeling for goral in Rajaji National Park (Kushwaha et al. 2000a, b), habitat evaluation for sarus crane in Keoladeo National Park, India (Palria et al. 2005) and rhino in Chitwan National Park, Nepal (Kafley et al. 2009) are noteworthy to mention.

This paper aims to investigate the spatial and temporal land cover changes in Kaziranga National Park and understand the possible causes of the changes. The study also attempts to analyze habitat condition to predict suitable habitat for rhinoceros in Kaziranga National Park.

10.2 Study Area

Kaziranga is located between latitudes 26°30′N and 26°45′N and longitudes 93°08′E to 93°36′E within Nagaon and Golaghat districts of Assam in the north-eastern part of India (Fig. 10.1). It lies on the flood plains of Brahmaputra River, sloping gradually from east to west against a backdrop of the foothills and snow-covered peaks of the eastern Himalayas. The park is a mosaic of eastern wet alluvial grasslands, alluvial plains, semi evergreen forests, tropical moist mixed deciduous forests, eastern Dillenia swamp forests, wetlands and sandy islands ('char'). The northern boundary of the park is marked by the Brahmaputra River while the hills of Karbi plateau lies towards south. Typical sub-tropical climate with temperature varies from 7° to 38 °C prevails in the park. Annual average rainfall is 1,320 mm.

There are four main types of vegetation: alluvial inundated grasslands and reed beds, alluvial savanna woodland, tropical moist mixed deciduous forests and tropical semi evergreen forests. Grasslands predominate in the western part of the park. Species like *Hemarthacompressa*, *Cynodondactylon*, *Cenchrusciliaris*, *Crysopogon aciculate* and *Andropogon* sp. dominate the short grasslands. Some of the important species of tall grasslands are *Phragmitiskarka*, *Saccharuamprocerum*, *S. spontanoum*, *Vertiveriazizaniodes*, *Thermedaarundinacea*, *Erianthusarundinacea and Imperatacylindrica*.

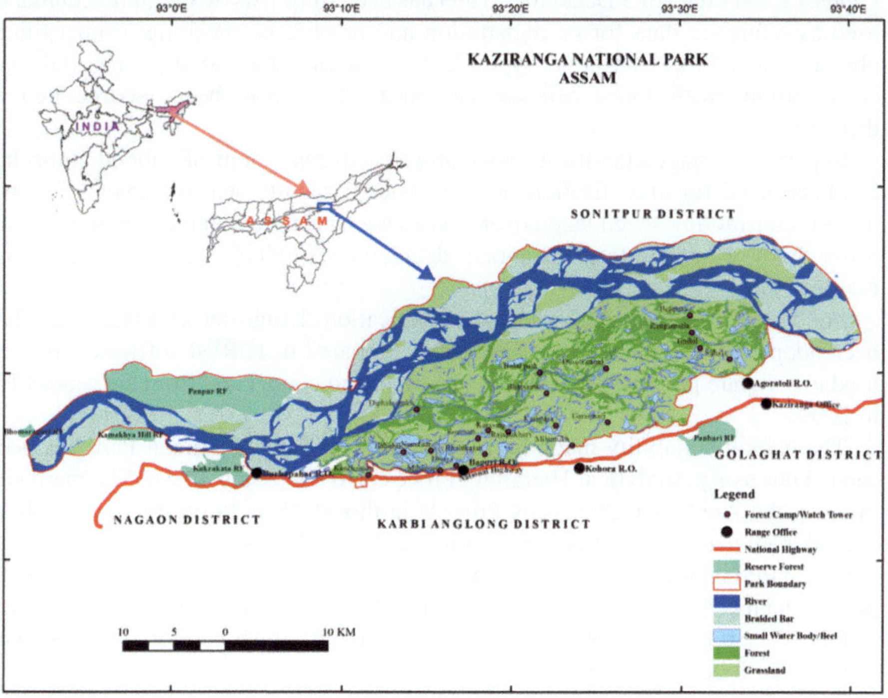

Fig. 10.1 Location map of the study area

In the tropical semi-evergreen forests common trees and shrubs are *Albiziaprocera, Duabangagrandiflora, Lagerstroemia speciosa, Cratevaunilocularis, Sterculiaurens, Grewiaserrulata, Mallotusphilippensis, Brideliaretusa, Aphaniarubra, Leeaindica and L. umbraculifera* (Kushwaha et al. 2000a, b, Patar 2005).

Apart from the one horned rhino, KNP is also famous for tiger (*Pantheratigris*), swamp deer (*Cervusduvauceliranjitsinhi*), hog deer (*Axis porcinus*), wild buffalo (*Bubalusbubalis*), elephants (*Elephasmaximus*). Other mammals include a small proportion of sambar (*Cervus unicolor*), barking deer (*Muntiacusmuntjak*), hoolock gibbon (*Bunopithecushoolock*), capped langur (*Trachypithecuspileatus*), jackal (*Canisaureus*), sloth bear, Ganges dolphin (*Platanistagangetica*), leopard, gaur (*Bosgaurus*) and otter.

10.3 Database and Methodology

In order to assess the changes in the national park, imageries of two time periods with a 20-year span period have been chosen for the study. Moderate resolution Indian Remote Sensing satellite data IRS LISS II (1990) and IRS LISS III (2009) of 36.25 m and 23.5 m spatial resolution respectively have been acquired from National Remote Sensing Center (NRSC), Hyderabad. In addition, ortho-corrected Landsat ETM+ data of 3 December, 1999 has been (http://www.glcf.umd.edu/data/) used as reference data for co-registration and resolution matching. Primary field observations of the land cover types, GPS locations of rhino sightings, informal conversations with forest officials and locals have also been incorporated in this work.

Supervised image classification technique with maximum likelihood algorithm has been used for classification of co-registered remote sensing images for two periods considering seven land cover classes viz. short grass, tall grass, river, small water body ('beel'), tropical moist deciduous, tropical semi evergreen and braided bars.

For change detection analysis, post-classification change detection technique has been adopted. The change detection matrix prepared in IDRISI software has been used to compare pixel by pixel changes in each land cover class from the years 1990 to 2009.

The habitat suitability mapping for rhino in Kaziranga National Park has been carried out using Analytical Hierarchy Process (AHP) (Saaty 1980). The main aim in using the Analytical Hierarchy Process in this study is to prioritize the habitat variables as per their preference. The hierarchy of the habitat variables has been created based on literature review on rhino, field observations and expert opinion of park officials. As the Analytical Hierarchy Process is based on nine point scale (Table 10.1), it is favourable for the study to rank the habitat variables of rhino such as grass for food, water, grass for shelter etc.

To perform habitat suitability mapping for rhino, four thematic data layers viz. land cover, proximity to road, proximity to water body and proximity to settlements

Table 10.1 Nine point scale of preference between two parameters in AHP (Saaty 1980)

Scales	Degrees of preferences
1	Equal importance
3	Moderate prevalence of one over another
5	Strong or essential prevalence
7	Very strong or demonstrated prevalence
9	Extremely high prevalence
2, 4, 6, 8	Intermediate values
Reciprocals	For inverse comparison

have been created using remote sensing and GIS for two time period (i.e. 1990 and 2009). The thematic layers have been generated using standard remote sensing-GIS procedure. By applying AHP method, the normalized weight/rating values of the classes/layers have been obtained (Table 10.2) and habitat suitability index have been calculated by using following equation:

$$\text{Habitat Suitability Index (HSI)} = \sum (\text{Rating} \times \text{Weighted thematic layer})$$

The procedure has been repeated for the two time period (i.e., 1990 and 2009) to obtain corresponding habitat suitability index maps, which has been reclassified into three equal interval classes, viz. highly suitable, moderate and less suitable zones for rhino. For validation of the rhino habitat variables GPS points of rhino sightings have been used.

10.4 Results and Discussion

10.4.1 Land Cover Mapping and Change Detection

Land cover maps of the park generated for 1990 and 2009 using maximum likelihood classification and validated through field check and high resolution Google Earth images are shown in Fig. 10.2. The distribution of various land cover classes for 1990 and 2009 is shown in Table 10.3.

The grassland areas cover a considerable area in the park among vegetation classes which is a prime habitat for rhino. Water bodies in Kaziranga National Park include Brahmaputra River and its three tributaries e.g. JiyaDiffolu, Mora Diffolu and Bhengra and the wetlands (locally known as ‘beels’). Mora Diffolu follows the southern boundary park and the Diffolu and Bhengra rivers flow through the park from east to west. There are almost 200 ‘beels’ distributed throughout the park.

For assessing the changes in the land cover in the park during 1990–2009, post-classification change detection matrix has been generated (Table 10.4). The change detection matrix shows spatial and temporal changes of 20 years span period in each land cover classes in the study area. Short grass areas have shown a positive increase (2 %). Increase in the short grass can be caused by conversion of ‘beels’ to

Table 10.2 Pairwise comparison of all Thematic layers (1990 and 2009)

Factors/Thematic layers	1	2	3	4	5	6	7	Weights
Data layers								
Land cover	1	5	8	9				0.640
Proximity to water	1/5	1	6	7				0.250
Proximity to road	1/7	1/6	1	2				0.066
Proximity to settlements	1/9	1/7	1/2	1				0.044
Land cover								
Short grass	1	3	3	5	5	7	9	0.371
Tall Grass	1/3	1	3	5	5	7	7	0.262
Small water body/beel	1/3	1/3	1	5	5	5	7	0.174
Tropical moist deciduous	1/5	1/5	1/5	1	1	3	3	0.065
Tropical semi evergreen	1/5	1/5	1/5	1	1	3	3	0.065
River	1/7	1/7	1/5	1/3	1/3	1	3	0.037
Braided bar	1/9	1/7	1/7	1/3	1/3	1/3	1	0.026
Proximity to water								
Small water body	1	2	3	5	7	7	9	0.363
0–250 m	1/2	1	2	3	5	7	9	0.249
250–500 m	1/3	1/2	1	2	3	5	9	0.161
500–1,000 m	1/5	1/3	1/2	1	2	3	9	0.102
1,000–2,000 m	1/7	1/7	1/3	1/2	1	2	9	0.066
Beyond 2,000 m	1/7	1/7	1/5	1/3	1/2	1	9	0.045
River	1/9	1/9	1/9	1/9	1/9	1/9	1	0.014
Proximity to road								
Beyond 100 m	1	3	5					0.637
50–100 m	1/3	1	3					0.258
0–50 m	1/5	1/3	1					0.105
Proximity to settlements								
Beyond 2,000 m	1	3	5					0.637
1,000–2,000 m	1/3	1	5					0.258
0–1,000 m	1/5	1/5	1					0.105

short grass. Around 0.38 % (2009) short grass area has been converted to 'beels'. The short grass and 'beel' areas are interchangeable to some extent because when the water level in the 'beels' recedes, the short grass area increases. Opposite trend is seen when water increases during the floods (Kushwaha et al. 2000a, b). On the other hand, tall grass has reduced by 6 % which indicate that overall grassland areas have been decreased. The conversion of tall grass to short grass (3.6 %) and 'beel' (1.3 %) could be considered as positive change. Reduction in the tall grasslands may be due to the expansion of forested areas into grasslands. Invasion of Mimosa in the short and tall grassland areas may also reduce the grassland areas in the park. Conversion of tall grass to tropical deciduous (0.9 %) and tropical semi evergreen (1.2 %) may also reduce the grassland areas. There has been increasein the forest classes tropical evergreen (4 %) and moist deciduous (0.7 %). Though the area covered by both the forest classes is less as compared to grassland areas the growth

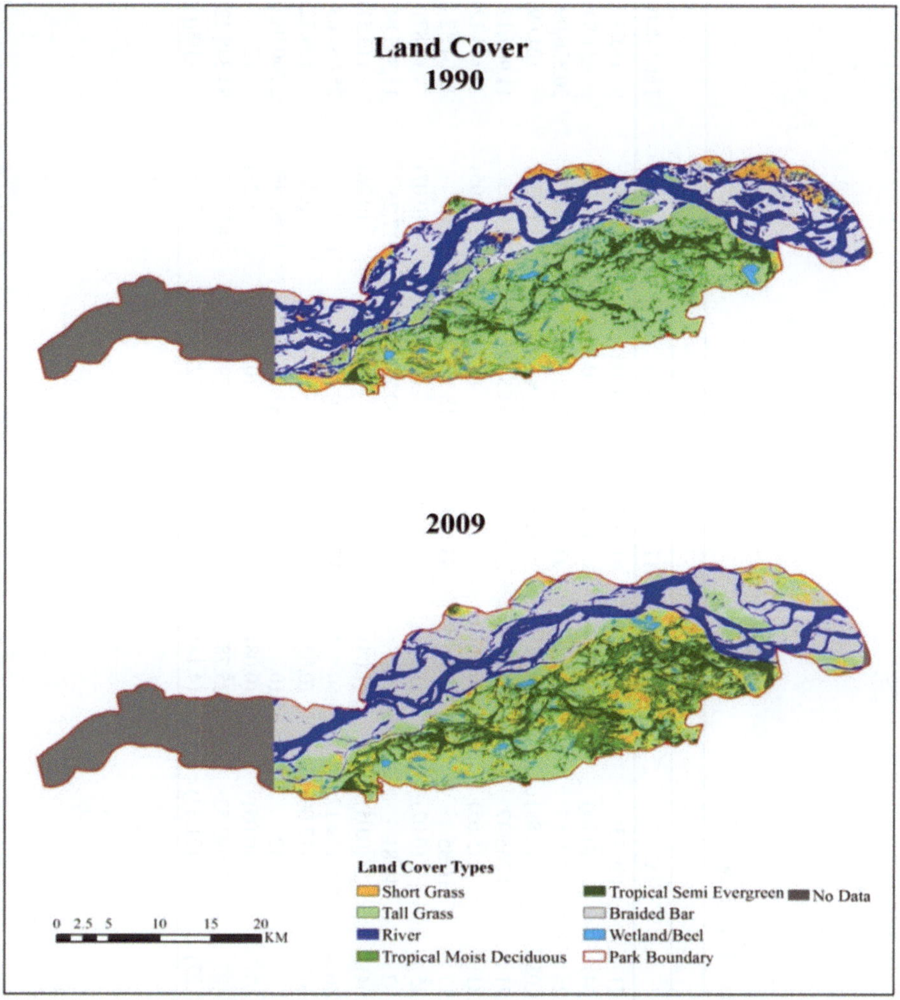

Fig. 10.2 Land cover map of 1990 and 2009 generated by image classification

Table 10.3 Distribution of land cover classes (1990 and 2009)

	1990		2009	
Land cover class	Area (ha)	Percentage	Area (ha)	Percentage
Short Grass	7,706.97	9.1	9,313.09	11.0
Tall Grass	25,529.03	30.3	20,320.09	24.1
River	17,993.52	21.3	11,907.17	14.1
Tropical Deciduous	3,525.78	4.2	4,104.32	4.9
Tropical Evergreen	7,899.83	9.4	11,395.84	13.5
Braided Bar	18,797.32	22.3	25,066.63	29.7
Wetland ('beel')	2,895.67	3.4	2,240.98	2.7
Total	84,348.12	100	84,348.12	100

Table 10.4 Change detection matrix 1990 (Columns)–2009 (Rows)

	1990 (ha)							
2009 (ha)	SG	TG	RV	TD	TE	BB	SW	Total (ha)
Short Grass (SG)	1,500.35 (1.78)	5,097.21 (6.04)	499.73 (0.59)	808.66 (0.96)	613.99 (0.73)	476.09 (0.56)	317.05 (0.38)	9,313.09 (11.04)
Tall Grass (TG)	3,044.39 (3.61)	9,343.24 (11.08)	2,430.67 (2.88)	819.65 (0.97)	973.06 (1.15)	2,579.89 (3.06)	1,129.19 (1.34)	20,320.09 (24.09)
River (RV)	457.15 (0.54)	1,524.71 (1.81)	4,765.37 (5.65)	46.89 (0.06)	221.01 (0.26)	4,852.95 (5.75)	39.10 (0.05)	11,907.17 (14.12)
Tropical Deciduous (TD)	253.54 (0.30)	2,069.17 (2.45)	16.35 (0.02)	847.70 (1.01)	836.27 (0.99)	18.28 (0.02)	63.01 (0.07)	4,104.32 (4.87)
Tropical Evergreen (TE)	533.53 (0.63)	4,758.02 (5.64)	40.81 (0.05)	881.28 (1.04)	4,934.19 (5.85)	30.76 (0.04)	217.26 (0.26)	11,395.84 (13.51)
Braided Bar (BB)	1,613.78 (1.91)	2,204.36 (2.61)	10,199.17 (12.09)	55.50 (0.07)	175.06 (0.21)	10,751.53 (12.75)	67.21 (0.08)	25,066.63 (29.72)
Small Water Body 'Beel' (SW)	304.23 (0.36)	532.31 (0.63)	41.42 (0.05)	66.10 (0.08)	146.24 (0.17)	87.81 (0.10)	1,062.86 (1.26)	2,240.98 (2.66)
1990 Total	7,706.97 (9.14)	25,529.03 (30.27)	17,993.52 (21.33)	3,525.78 (4.18)	7,899.83 (9.37)	18,797.32 (22.29)	2,895.67 (3.43)	84,348.12 (100)

Area in percentages is given within parenthesis

rate of the area is more than the grasslands. Both the water body class river and 'beel' have also been decreased. River area covered 21 % during 1990, however, it has been reduced to 14 % by 2009. Decline in river area is attributed to erosion due to annual devastating floods. During 1990 'beel' occupied 3.4 % which decreased to 2.6 % by 2009 showing overall decrease in the area. 'Beel' area being converted to short grass (0.4 %) and tall grass (0.6 %) can increase the grassland areas for rhino. However, increasing sedimentation or silt deposition by recurrent flood water in the water bodies has caused reduction in the area and degradation in the quality of 'beels'. Some amount of grasses growing in the silted areas are not of good quality, hence, herbivores do not feed on these grasses. As a matter of fact, siltation of the water bodies is one of the vital factors towards habitat degradation in the park. 'Beel' are important for many species such as rhino, wild buffalo, swamp deer, birds etc. Therefore reduction in the 'beels' may affect the fauna. The study findings reflect the dynamics of Kaziranga topography. Reduction in the grassland and wetlands can be a major threat to the rhino habitat as grass and water are the most important habitat requirement for rhino.

10.4.2 Habitat Change Mapping

Habitat suitability mapping for rhino has been done to obtain the suitable areas in the park. As per literature review and field observations, grassland and 'beels' have been considered as the major habitat variables for rhino. Rhinos prefer grasslands to forest areas as their diet strongly dominated by grass both short and tall grass. Wallowing is an important activity of the rhinos. 'Beels' have been used by rhinos for wallowing to release excess heat and to protect themselves from insects. The rhinos go to the forested areas during the time of flood when majority of the grassland areas get submerged. The habitat suitability maps (Fig. 10.3) based on AHP technique showed that highly suitable areas occupy the areas of short grass tall grass and 'beels'. Moderately suitable area falls in the areas of forests and some part of sandy braided bar area. Low suitable areas cover the Brahmaputra River on the northern boundary of the park and some parts in the north eastern area of the park.

As in land cover change detection procedure, a similar approach has been adopted to quantify the change in habitat in two time periods with the help of habitat change detection matrix (Table 10.5). It is found that high suitable areas have been reduced by 7 % from 1990 to 2009. On the other hand, low and moderately suitable areas have been increased by 4 % and 3 % percent from 1990 to 2009 respectively. Highly suitable areas converted to moderate (10.6 %) and low (7.6 %) suitable areas could apparently cause decline in the habitat for rhino. It can also be seen that the area covered by highly suitable areas (40.6 %) during 1990 has been reduced to 33.6 % by 2009. During 1990, highly suitable areas covered maximum area of the park. However, 2009 scenario of suitability shows that the area covered by highly suitable areas has been decreased. Land cover change in the park during this period is apparently the reason for change and conversion to other suitability classes as the

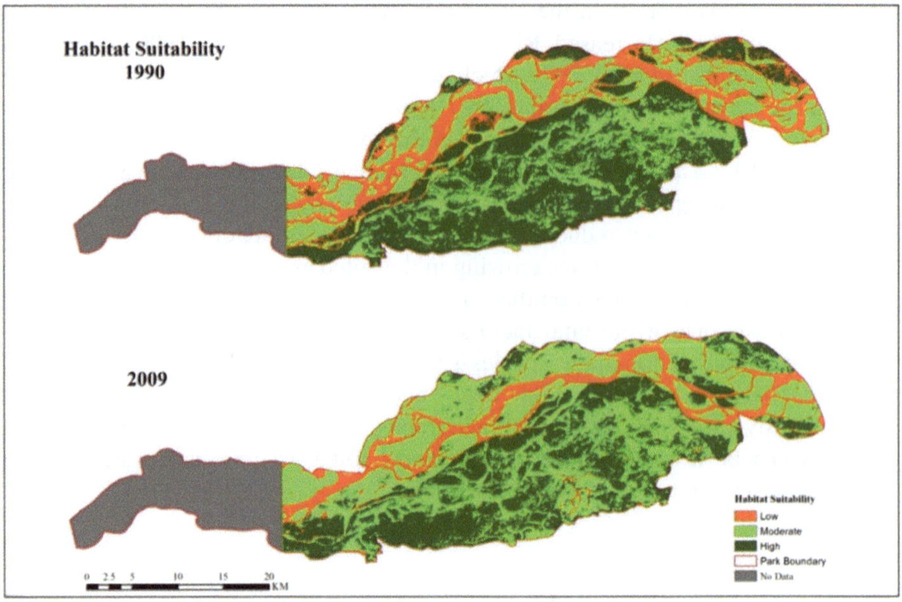

Fig. 10.3 Habitat suitability maps (1990 and 2009) generated by AHP

Table 10.5 Habitat change matrix (in percentage): 1990 (Columns)–2009 (Rows)

	1990			
2009	Low	Moderate	High	Total
Low	17.78	10.91	7.63	36.33
Moderate	9.13	10.28	10.64	30.05
High	5.17	6.06	22.39	33.62
Total	32.08	27.25	40.66	100

land cover classes have been used as variables for obtaining the habitat suitability index. Hence, the main reasons for change in the land cover can be associated with the changes in suitable habitats for rhino in the park.

10.4.3 Impact on Rhino Habitat

Previous studies on changes in land cover in wildlife areas stated that disturbance in land cover can affect the wildlife habitat (Townsend et al. 2009). Evidently, land cover change in any protected area would have some effect on wildlife habitat.

The change detection analysis for two periods has shown significant changes in vegetation and wetland area. From the habitat suitability analysis, it is observed that short grass, tall grass and 'beel' classes are considered suitable for rhino

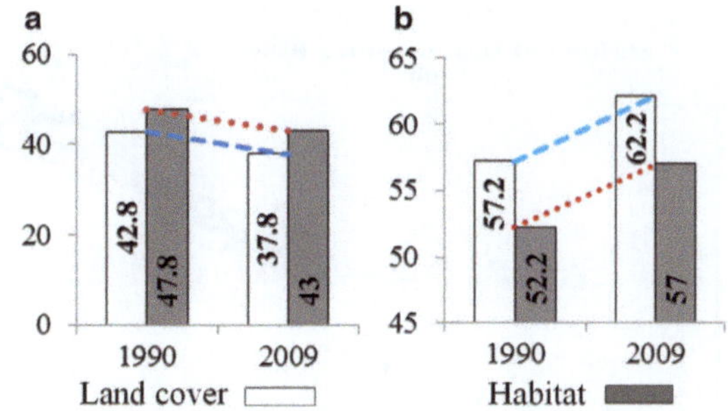

Fig. 10.4 (**a**) Suitable and (**b**) Non suitable land cover and habitat for rhino (1990 and 2009)

habitat, whereas forest classes tropical moist deciduous forest and tropical semi evergreen forest along with river and braided bar class are considered not suitable. Therefore, increase or decrease in land cover would show change in the suitable habitat for rhino.

Comparison has been done between the land cover classes and habitat suitability classes. The comparison has showed that with the decrease in suitable land cover, suitable habitat areas for rhino have also shown decline. Simultaneously, increase in unsuitable land cover areas has shown increase in unsuitable habitat areas from 1990 to 2009 (Fig. 10.4a, b). It can be derived that any change in the land cover would show change in the rhino habitat. The comparative analysis has proved that there exists the impact of land cover change on the rhino habitat change [Medhi A (2011) Impact of landuse/cover change on the rhino habitat in Kaziranga National Park, Assam. M.Phil dissertation, University of Delhi (Unpublished)].

Flooding is the causative factor for creation, maintenance and eradication of moist alluvial grasslands. Floods occur every year, sometimes several times in a year. Brahmaputra River is the immediate boundary of the park. During floods the excess water submerges the area and covers 80–90 % of the total land area. During high floods especially when the 'beels' and grasslands are submerged, the animals from the Park suffer from shortage of fodder and often migrate to the neighbouring hills of Karbi-Anglong across the NH-37.

It has been observed that the rivers, streams and 'beels' are getting silted as a result of which there is a reduction in the size of the wetlands and moist alluvial grassland which is regarded as a highly suitable for rhinos. There are quite a number of 'chapories' (braided bars) of various sizes which have been formed by silt depositions and confirms with our finding of 7 % increase in such land cover (Table 10.4). It is observed that large scale accretion is taking place near the north-western boundary of the park. From 1967 to 2008, accretion of about 3,329.46 ha occurred in Kaziranga National Park areas (GOA (Government of Assam) 2009).

The northern boundary of the park formed by the Brahmaputra River seems to be highly unstable. Shifting of the Brahmaputra River towards south is a serious matter of concern (Fig. 10.5). Erosion has been a major threat to the Park as it has already lost considerable area within it due to erosion by the Brahmaputra River every year. From 1990 to 2009 considerable amount of area has been lost due to erosion. About

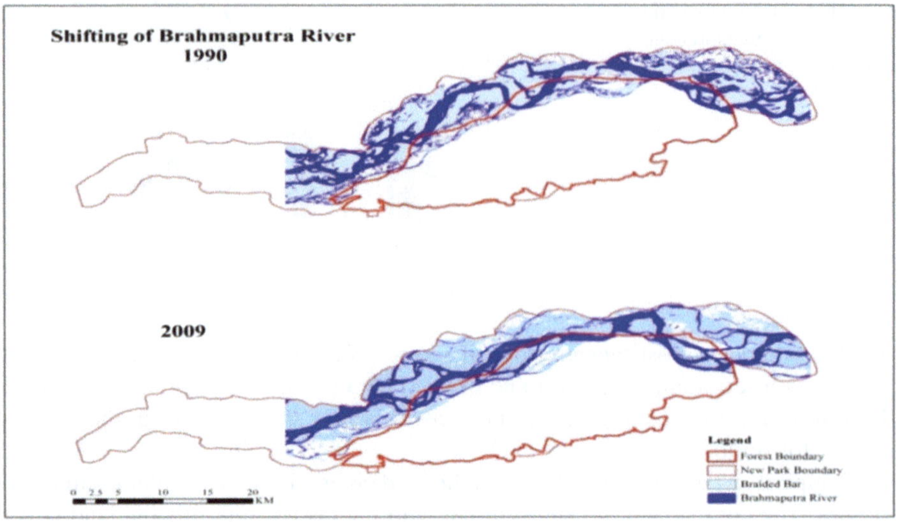

Fig. 10.5 Shifting of Brahmaputra river (1990 and 2009)

1,063 ha of land area has been eroded from 1990 to 2009 (Fig. 10.5). Reduction the park area may affect the habitat for many species in the park. However, this can be an alarming sign for conservation management of the park in future.

Despite the devastating annual floods and high rate of poaching and land cover change inside the park, the rhino population has not been reduced. The population census of rhino shows that the rhino population is at a better state and increasing per year which is a remarkable achievement of the park. According to an IUCN report (Foose and Strien 1997), the rhino carrying capacity in Kaziranga is 1,500. However, Kaziranga during 1999 has exceeded the carrying capacity. The rhino population has reached to 2048 during the 2009 census which is more than 85 % of the one horned rhino population present in the world. According to 2012 census, rhino population has increased to 2,290. The conservation management effort is not only for the increase of the rhino population but also managing their habitat. Therefore, it is really necessary to maintain their habitat so that the rhino population can comfortably live in these areas.

Hence, habitat availability for rhino is also essential to for conservation. The suitable area available for each rhino has been calculated using the area total suitable area of the park divided by number of rhino in a particular year. Figure 10.6a shows the suitable area available for each rhino in the national park. The area available for each rhino during 1990 was approximately 0.31 km^2. The area has been decreased to 0.16 km^2 during 2009. From this, it is clear that, year by year the rhinos are increasing because of the presence of suitable habitat and management effort. But simultaneously, the suitable area available for each rhino is decreasing in Kaziranga. There are also other herbivores such as wild buffalo and swamp deer which can give competition to rhino for food (Fig. 10.6b). The growing

Fig. 10.6 (a) Area available for each rhino (km^2) and (b) Area available for each rhino, swamp deer and wild buffalo (km^2)

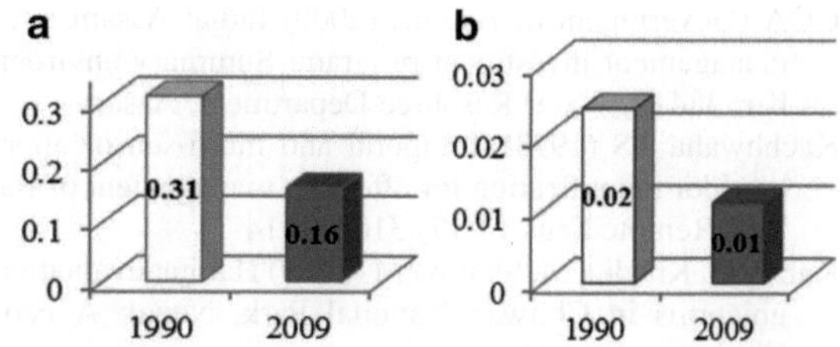

population of wild buffalos and swamp deer is also a matter of concern. The feeding behaviour of these two herbivores is similar with rhino. Wild buffalo is voracious eater. It is really making pressure on the food for rhino. The grasslands are overexploited. Therefore, increasing population of these two herbivores may increase the competition of food for rhino [Medhi A (2011) Impact of landuse/ cover change on the rhino habitat in Kaziranga National Park, Assam. M.Phil dissertation, University of Delhi (Unpublished)]. The ecological studies on the other herbivores would be helpful for managing the park effectively.

10.5 Conclusions

The above discussion provided a clear picture of the relationship between land cover change and habitat suitability. Comparative analysis of land cover and habitat suitability change highlighted the impact of land cover on habitat suitability change. Linking the rhino population with available areas for rhino showed decline in the available areas for each rhino with the competition from other herbivores. Increasing rhino population in the reduced park area needs proper management including various conservation measures, such as, translocation to some other suitable habitats and creation of wildlife corridor between two protected areas. The study can be considered significant from the conservation point of view for providing necessary database on land cover trends of Kaziranga National Park as well as indicating suitable areas for rhino habitat. The study has also proved the importance of application of remote sensing and GIS technologies in monitoring national parks.

References

Bayarsaikhan U, Boldgiv B, Kim KR, Park KA, Lee D (2009) Change detection and classification of land cover at Hustai National Park in Mongolia. Int J Appl Earth Obs Geoinf 11:273–280
Foose TJ, Strien NV (eds) (1997) Asian Rhinos. IUCN, Gland, Switzerland and Cambridge
Gross JE, Goetz SJ, Cihlar J (2009) Application of remote sensing to parks and protected area monitoring: introduction to the special issue. Remote Sens Environ 113:1343–1345

GOA (Government of Assam) (2009) India: Assam integrated flood and riverbank erosion risk management investment program: Summary environmental impact assessment project number: 38412, Water Resource Department, Assam

Kachhwaha TS (1993) Temporal and multi-sensor approach in forest vegetation mapping and corridor identification for effective management of Rajaji National Park, Uttar Pradesh, India. Int J Remote Sens 14(17):3105–3114

Kafley H, Khadka M, Sharma M (2009) Habitat evaluation and suitability modeling of Rhinoceros unicornis in Chitwan National Park, Nepal: A geospatial approach, XIII World Forestry Congress, Argentina

Kennedy RE, Townsend PA, Gross JE, Cohen WB, Bolstad P, Wang PC, Adams P (2009) Remote sensing change detection tools for natural resource managers: understanding concepts and tradeoffs in the design of landscape monitoring projects. Remote Sens Environ 113:1382–1396

Kushwaha SPS, Unni NVM (1986) Applications of remote sensing techniques in forest cover monitoring and habitat evaluation- A case study at Kaziranga National Park, Assam. In: Kamat DS, Panwar HS (eds) Proceedings of seminar-cum-workshop on wildlife habitat evaluation using remote sensing techniques. Indian Institute of Remote Sensing/Wildlife Institute of India, Dehradun, pp 238–247

Kushwaha SPS, Munkhtuya S, Roy PS (2000a) Geospatial modeling for goral habitat evaluation. J Indian Soc Remote Sens 28(4):293–303

Kushwaha SPS, Roy PS, Azeem A, Boruah P, Lahon P (2000b) Land area change and rhino habitat suitability analysis in Kaziranga National Park, Assam. Tiger Paper XXVIII(2):9–16

Nagendra H, Tucker C, Carlson L, Southworth J, Karmacharya M, Karna B (2004) Monitoring parks through remote sensing: Studies in Nepal and Honduras. Environ Manage 34(5):748–760

Pant A, ChavanSG RPS, Das KK (1999) Habitat analysis for sambar in Corbett National Park using remote sensing and GIS. J Indian Soc Remote Sens 27(3):133–139

Palria S, Singh A, Sharma JR, Pathak S (2005) Habitat evaluation for sarus crane in the Keoladeo National Park using IRS LISS III and PAN merged data and GIS. J Indian Soc Remote Sens 33 (2):259–266

Parihar JS, Panigrahy S, Lahan P (1986) Remote sensing based habitat assessment of Kaziranga National Park. In: Kamat DS, Panwar HS (eds) Proceedings of seminar-cum-workshop on wildlife habitat evaluation using remote sensing techniques. Indian Institute of Remote Sensing/Wildlife Institute of India, Dehradun, pp 157–164

Patar KC (2005) Behavioural patterns of the one horned Indian Rhinoceros. Spectrum Publications, India

Sarma PK, Lahkar BP, Ghosh S, Rabha A, Das JP, Nath NK, Dey S, Brahma N (2008) Land-use and land-cover change and future implication analysis in Manas National Park, India using multi-temporal satellite data. Curr Sci 95(2):223–227

Saaty TL (1980) The analytical hierarchy process. McGraw-Hill, New York

Saxena KG (1986) Forest cover changes between proposed Rajaji National Park and Corbett National Park during the period 1972–1983 for identifying elephant corridors. In: Kamat DS, Panwar HS (eds) In: Proceedings of the seminar-cum-workshop on wildlife habitat evaluation using remote sensing techniques, October 22–23 1986, pp 229–237

Townsend PA, Lookingbill TR, Kingdon CC, Gardner RH (2009) Spatial pattern analysis for monitoring protected areas. Remote Sens Environ 113:1410–1420

Wang Y, Mitchell BR, Nugranad-Marzilli J, Bonynge G, Zhou Y, Shriver G (2009) Remote sensing of land-cover change and landscape context of the National Parks: a case study of the Northeast Temperate Network. Remote Sens Environ 113:1453–1461

Chapter 11
Land Use and Land Cover Change Along Shivaliks Between River Ghagghar and Yamuna

Dimple and Ravinder Singh

Abstract In the past two centuries the impact of human activities on the land has grown enormously, altering almost entire landscapes, and ultimately impacting the earth's nutrient and hydrological cycles as well as climate. Humans have been altering land cover since pre-history, through the use of fire to flush out game and, since the advent of plant and animal domestication, through the clearance of patches of land for agriculture and livestock. Land-use and land-cover changes are local and place specific, occurring incrementally in ways that often escape our attention. Yet, collectively, they add up to one of the most important facets of global environmental change. The present study is intended to study the area along the Shivaliks hills, between river Ghagghar and Yamuna. The developmental blocks of the foothill zone of Shivaliks bordering Haryana have been taken for the study. The objective of the study is to study the patterns of land use and land cover in the area during year 1990, 2000 and 2010 detect the changes in the land use and land cover pattern during this period and to identify and investigate the factors responsible for the change. The satellite based remotely sensed data is used to collect the land use and land cover pattern and the analysis of the data in done in GIS software. The results portraits that there remain only few landscapes in the study area those are still in their natural state. Due to anthropogenic activities, the study region is being significantly altered in some manner and man's presence on the area, his use of land has had a profound effect upon the natural environment thus resulting into an observable pattern in the land use/land cover over time. The land use/land cover pattern of a region is an outcome of natural and socio–economic factors and their utilization by man in time and space. The main objective of the study was to test the remote sensing and GIS for the purpose of detection of change in land use and land cover. When the results from remote sensing and GIS were compared with the

---

Dimple • R. Singh (✉)
Research Scholars (JRF), Department of Geography, Panjab University, Chandigarh, India
e-mail: Dimpalsinghal01@gmail.com; Ravinder3080@gmail.com

M. Singh et al. (eds.), *Climate Change and Biodiversity: Proceedings of IGU Rohtak Conference, Vol. 1*, Advances in Geographical and Environmental Sciences,
DOI 10.1007/978-4-431-54838-6_11, © Springer Japan 2014

ground reality they are found quite reliable. The results are found near to the reality. Different time period satellite imageries are very useful for the detection of changes. This method saves efforts and time of the researcher who wants to investigate the changes taking place over an area.

Keywords GIS • Land use and land cover change • Remote sensing

11.1 Introduction

Humans have been altering land cover since pre-history through the use of fire to flush out game and, since the advent of plant and animal domestication, through the clearance of patches of land for agriculture and livestock. In the past two centuries the impact of human activities on the land has grown enormously, altering entire landscapes, and ultimately impacting the earth's nutrient and hydrological cycles as well as climate. Land-use and land-cover changes are local and place specific, occurring incrementally in ways that often escape our attention. Yet, collectively, they add up to one of the most important facets of global environmental change. Land is one of the most basic and important natural resource (Petrson 2006). Man is directly or indirectly dependent on land for his basic needs and requirements, but from the last few years land is under serious pressure because of increasing population and increasing demands, needs and requirements of growing population.

11.2 Definitions and Concepts

11.2.1 Land

Land can be variously defined. Some of the important definitions of land are as follows:- "Land is the basic natural resources... it is perhaps regarded as a resource base rather than a resource itself" (Mather 1986). Any portion (large or small) of the surface of the earth, considered by itself, or as belonging to an individual or a people, as a country, estate, farm, or tract.

11.2.2 Land Use

Land use is the human use of land. Land use involves the management and modification of natural environment or wilderness into built environment such as fields, pastures, and settlements. It has also been defined as "the arrangements, activities and inputs people undertake in a certain land cover type to produce, change or maintain it" (FAO/UNEP 1997; FAO/UNEP 1999).

11.2.3 Land Cover

is the observed biophysical cover of the earth's surface. Strictly speaking, it describes vegetation and man-made features, and omits bare rock and water, although, in practice, these elements are often included under this term (Khullar 2005). Land cover is the physical material at the surface of the earth. Land covers include grass, trees, bare ground, water, etc.

11.2.4 Land-Use and Land Cover Change (LULCC)

Land-use and land-cover change studies provide valuable information for large-scale vegetation biomass and forest cover assessments that are key components of the carbon cycle. (Anderson et al. 1976) Future land-use and land-cover change goals include:

- Very accurate biomass estimates, thus refining knowledge of carbon storage in vegetation,
- Understanding regional land-use changes that affect biomass, and
- Quantifying linkages and feedbacks between land-use and land-cover change, climate change forecasting, climate change, and other related human and environmental components.

11.3 Role of Remote Sensing and GIS in Detection of Land Use and Land Cover Change

Remote sensing is the acquisition of information about an object or phenomenon, without making physical contact with the object. In modern usage, the term generally refers to the use of aerial sensor technologies to detect and classify objects on Earth. The science and art of obtaining useful information about an object, area or phenomenon through the analysis of data acquired by a device that is not in contact with the object, area, or phenomenon under investigation (Meyer 1995). GIS is a system of hardware and software used for storage, retrieval, mapping, and analysis of geographic data. Practitioners also regard the total GIS as including the operating personnel and the data that go into the system. Spatial features are stored in a coordinate system (latitude/longitude, state plane, UTM, etc.), which references a particular place on the earth. Descriptive attributes in tabular form are associated with spatial features. Spatial data and associated attributes in the same coordinate system can then be layered together for mapping and analysis (Pandy and Nathawat 2006). GIS can be used for scientific investigations, resource management, and development planning. In land use and land cover change studies too remote sensing and GIS has proved as most desirable tools for studying.

11.4 Objectives of the Study

Behind every research there are some objectives. Without objects no research works are conducted, the objectives behind the study of Land use and land cover changes along Shivaliks between river Ghagghar and Yamuna are as follows: First, to study the patterns of land use and land cover in the study area during the year 1990, 2000 and 2010. Second, to study the changes in land use and land cover from 1990 to 2010. And last, to identify and investigate the factors responsible for the change.

11.5 Study Area

The area along the Shivaliks hills is selected as the study area. The area between river Ghagghar and river Yamuna is taken as the study area. The Community Development (CD) blocks of the foothill zone of Shivaliks bordering Haryana have been taken as study area. Study area in this report includes the District Panchkula, Ambala and Yamunanagar and the corresponding CD blocks are:- Morni, Raipura rani, Narayangarh, Sadaura, Bilaspur, Chhachhrauli. The areal extent is from $30 \times 45'23.368''$ N to $30 \times 7'37.194''$ N and $77 \times 36'25.975''$ E to $76 \times 55'34.252''$ E. The state boundary of Uttar Pradesh and Uttrakhand touches the study area from North–Eastern side (Fig. 11.1)

11.6 Methodology

Methodology involves all those steps which are involved in designing of the research process. Following steps are involved in the present study (Fig. 11.2).

11.6.1 Data Acquisition

To detect the land use and land cover satellite imageries of minimum two different time periods are required. So, satellite imageries from Landsat ETM and IRS P6 LISS-III imageries area used. The Landsat Imagery of this area was downloaded from internet from http://landsat.usgs.gov/ website. The IRS P6 imagery was provided to Department of Geography, Panjab University, Chandigarh by NRSC. In addition Survey of India toposheets that cover the whole area are used. Further verification was conducted with the help of Google Earth to verify the signatures and to verify the results.

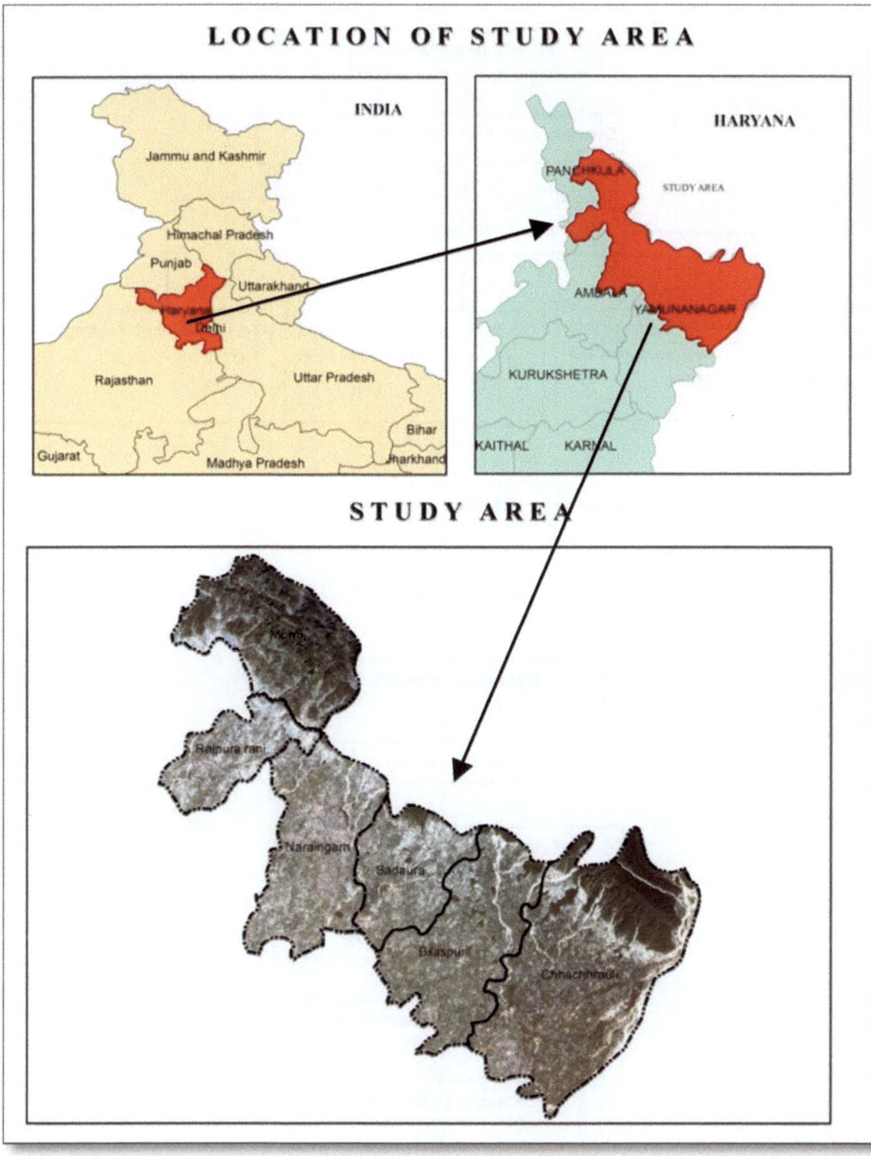

Fig. 11.1 Location of study area. *Source*: IRS, P6 Satellite Data & Census of India, Administrative Atlas of Haryana, 2001

11.6.2 Data Processing

For data processing visual image interpretation and digital image interpretation techniques are used. Signatures are identified from satellite imageries. Erdas Imagine 9.1 and ArcGIS 9.3 GIS software were used for data processing.

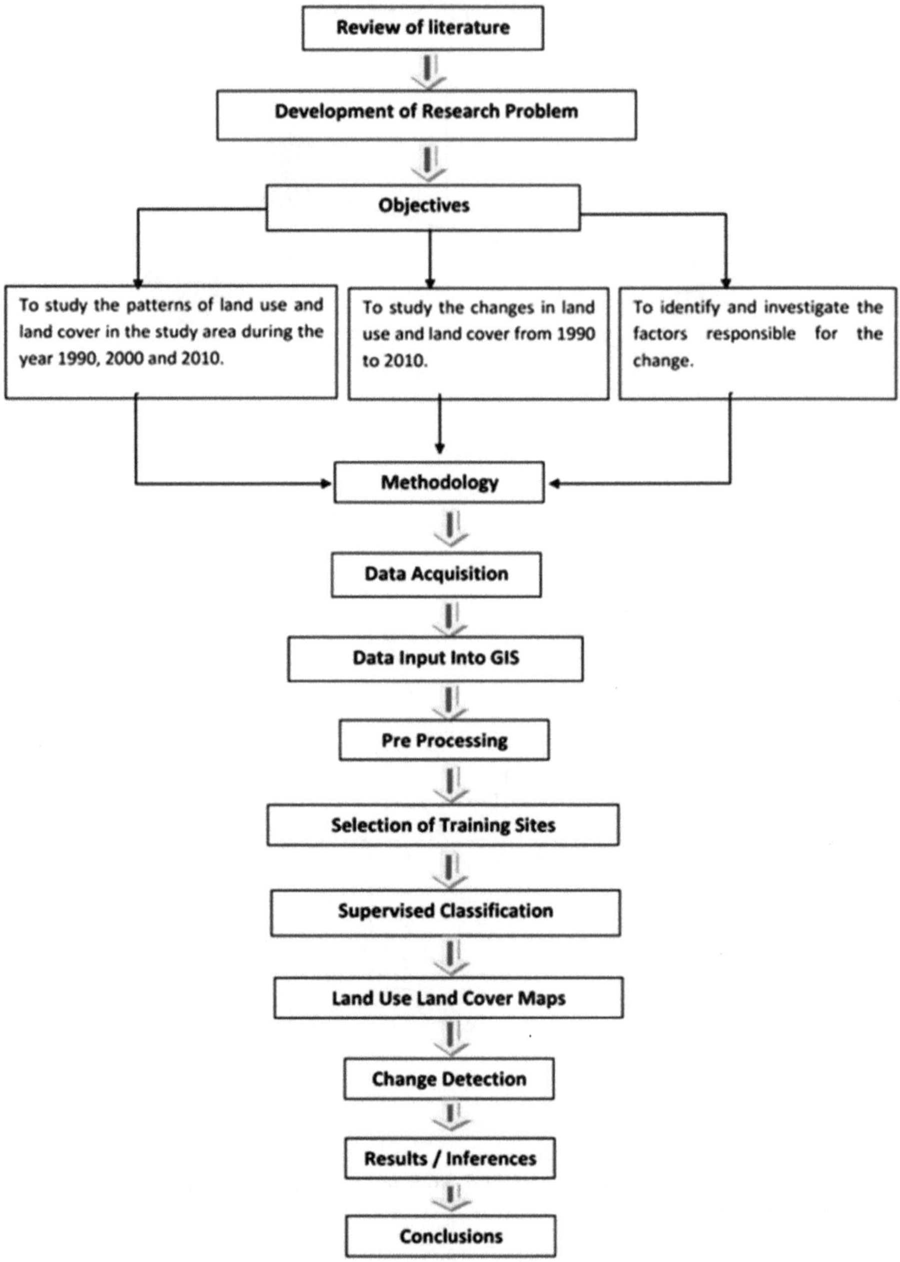

Fig. 11.2 Methodology of the study

Erdas Imagine 9.1 is perfect software for digital image interpretation. Supervised classification was done for all images. Signatures of the classes are identified from satellite imageries and verified in the field and on Google Earth. Eight classes of the land use and land cover are created in both time periods.

11.6.3 Data Analysis

The data which is processed with GIS softwares is also analyzed with these softwares. The changes in area under each category was detected, classified and then mapped (Robinson et al. 1995). All Practical exercise is done with GIS software's. These softwares are specially designed for these purposes. Mainly Erdas Imagine 9.1 is used for analysing the raster data.

11.6.4 Classes of the Land Use and Land Cover

For identifying the land use and land cover classes the scheme classification scheme developed by NRSA in 1995 is used. So on the basis of this Scheme 8 land use and land cover classes are created. These land use and land cover classes are:

1. Forest
2. Agricultural Land
3. Barren Land
4. Settlements
5. Water Bodies
6. Choes

The details of land use and land cover classification scheme developed by NRSA in 1995 is given in following table: (Table 11.1).

11.7 Land Use and Land Cover Pattern, 1990

A forest will be termed as dense forest when its canopy cover will be more than 60 % and it was observed that the total area under the dense forest in that year was 35,711.62 ha. And they are found mainly on the slopes of the Shivaliks. Some patches of dense forests were also found in central part of the study area. When the canopy cover is less than 60 % then it'll be termed as open forest and it was found that the total area under open forest in that year was 1,373.110 ha (Fig. 11.3).

Open forest mainly found in the linear patches in south-eastern part of the study area, but those patches were also sparsely distributed in the study region. The open forest are relatively less dense from dense forest. The open forest are located

Table 11.1 Land use and land cover classification categories

Level-I	Level-II	Level-III
1. Built up land	1.1 Towns/cities Villages	
	1.2 Villages	
2. Agricultural land	2.1 Crop land	2.1.1 Kharif
	2.2 Fallow Land	2.1.2 Rabi
	2.3 Plantation	2.1.3 Kharif + Rabi (Double Crop)
3. Forest	3.1 Evergreen/Semi evergreen	3.1.1 Dense
	3.2 Deciduous	3.1.2 Open
	3.3 Scrub Forest	
	3.4 Forest Blank	
	3.5 Plantation	
	3.6 Mangrove	
4. Wasteland	4.1 Salt affected land	
	4.2 Water logged land	
	4.3 Marshy/swampy	
	4.4 Land with Scrub	
	4.5 Sandy	
	4.6 Mining/Industrial waste	
	4.7 Barren Rocky/stony/waste/sheet rock area	
5. Water bodies	5.1 River/Stream	
	5.2 Canals	
	5.3 Lake/reservoirs/tank	
6. Others	6.1 Shifting cultivation	
	6.2 Grass land/grazing land	
	6.3 Salt pans	
	6.4 Snow covered	

Source: NRSC, LULC Classification, 1995

basically along the water bodies in the study area. The area along Shivaliks between the river Ghagghar and Yamuna is suitable for cultivation that is why there are very small patches of open forest in the present study area (Table 11.2).

Study have shown that the total area under settlement in the year 1990 was 1,874.16 ha. The size of the settlements were small and they were sparsely distributed all over the study area, except in the extreme north region (Fig. 11.4).

The land use patterns is showing that the settlements are dispersed all over the study area. The basic resion behind the sparsely distributed settlements is that the area is very good for the purpose of cultivation and agricultural activities. That is why most of the area is under the agricultural land use the, availability of water, good agricultural soils, and availability of amenities and facilities all over the area and at equal cost allows the people to distribute all over the study area. Only few water bodies are identified in this area. The total area under waterbodies was found to be 17.92 ha in the year 1990. No large water body is found in this area from the satellite imagery of this time period. Although the region is known for the very

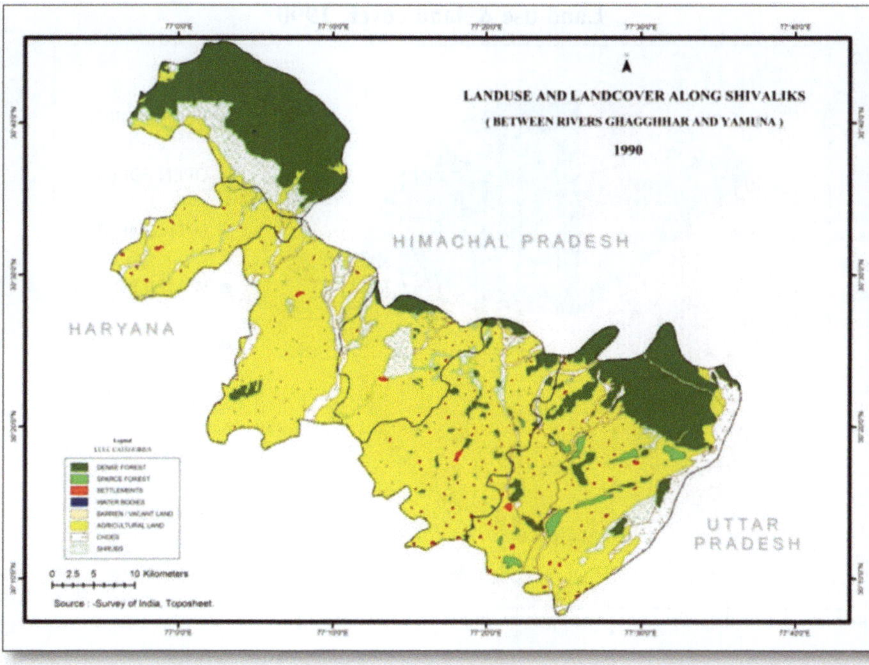

Fig. 11.3 Land use and land cover, 1990. *Source*: Survey of India, Toposheet, 1990

Table 11.2 Land use and land cover, 1990

S. No	Category	Area in hectares
1	Dense forest	35,711.62
2	Open forest	1,373.10
3	Settlement	1,874.16
4	Water bodies	17.92
5	Barren land	90.10
6	Agricultural land	105,395.6827
7	Choes	11,137.49
8	Scrubs	12,289.92
Total		167,890.00

Source: Spatial Analysis with ARC GIS from SOI Toposheet, 1990

good availability of the water resources but most of the water resources used in the region for the purpose of irrigation and drinking are under ground water resources. The major portion of the study area was irrigated by tube wells and canals that' why there are very few surface water sources. Studies have shown that there were some area under the barren land. Barren land is that land which is lacking vegetation, especially useful vegetation an uninhabited wilderness that is worthless for cultivation, and the area under barren land was 90.10 ha during the year 1990. In the year 1990 there were very few barren land as the category of barren land is very

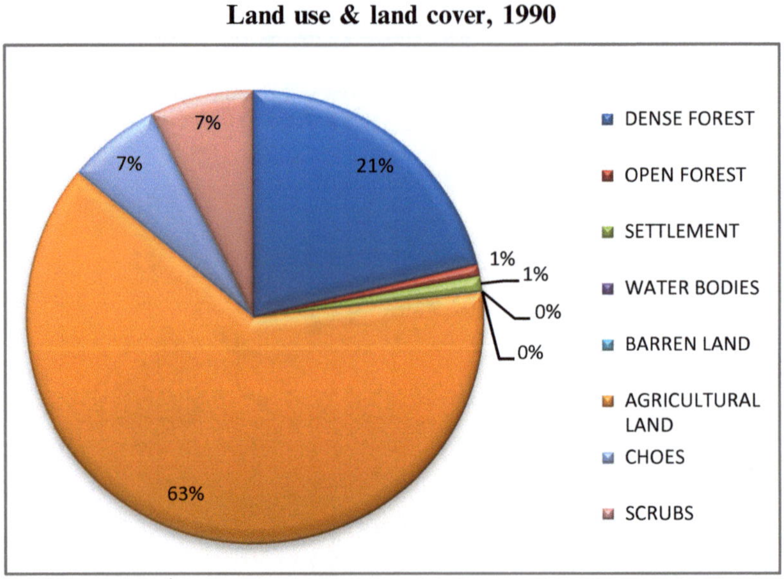

Fig. 11.4 Land use and land cover, 1990. *Source*: Table No. 11.2

close to be classified in scrubs. Initially there were very small area where the land was barren. It was after the mechanisation of the agricultural activities that the category of scrubs and choes were converted into agricultural land. Before including into the agricultural land the area left vacant was considered as barren because no agricultural activity was done on the area. In 1990 large part of the study region was occupied by the agricultural land and it was found that the total area under agricultural land was 105,395.6827 ha and it was the largest area under land use category of the year 1990. All the region is very favourable for the purpose of agriculture. The climatic conditions allow most of the region to be used as agricultural land. The soil is alluvial in most of the area which is very suitable for agricultural activities. The population of the area is dependent on the agriculture so they use maximum area for the purpose of agriculture. The agricultural productivity is major determinent of the agricultural area in the present study area. The total area under the choes was found to be 11,137.49 ha. In the year 1990. The choes are the prominent feature of teh kandi area of the foothill zone of shivalik range. The choes are the seasonal streams of the kandi area. They come into action in the monsoon season. The choes are basically very small in length which occupy basically very small distance from the foot hill region. In all parts of the study area choes can be seen. The choes are basically dry patches of the stream bed. They have sandy base. The choes occupy large proportion of the present study area. Some parts of the study region were under Scrubs. Scrubs are multi-stemmed woody plants with several stems that originate at the base of the plant. The natural shape of the Scrubs will vary according to their individual branching patterns and it was observed that the total area under the Scrubs was 12,289.92 ha.

11.8 Land Use and Land Cover Pattern, 2000

Dense forests are an important feature of the study area. They occupy 20,862.92 ha of the total area i.e. 12 % of the total area. Dense forests are in the north and north-eastern area in small patches.

They are found on the slopes of the hills. The total area under open forests is 19,636.56 ha. i.e. 12 % of the total area. The open forests are found in northern, north-eastern and south-eastern parts of the study area. Open forests also found in small patches in central part (Fig. 11.5, Table 11.3).

The study area is dominated by the agricultural land. The agricultural land in this area occupies 106,108.95 ha. i.e. 63 % of the total area. Except the hilly area, the whole area is under agriculture. Barren land is also found in this area, small dispersed signatures of barren land are identified on the satellite imagery. It constitutes 4,719.40 ha area i.e. 3 % of the total area. Small size settlements are also found in this area. They are dispersed in the agricultural fields in small clusters.

Most of the settlements are found in southern edge of the study region. The settlements occupy 7,918.62 i.e. 5 % of the total area. Several water bodies are identified in this area. This class occupies 580.02 ha i.e. less than 1 % of the total area. Several important water bodies are found in this area from the satellite

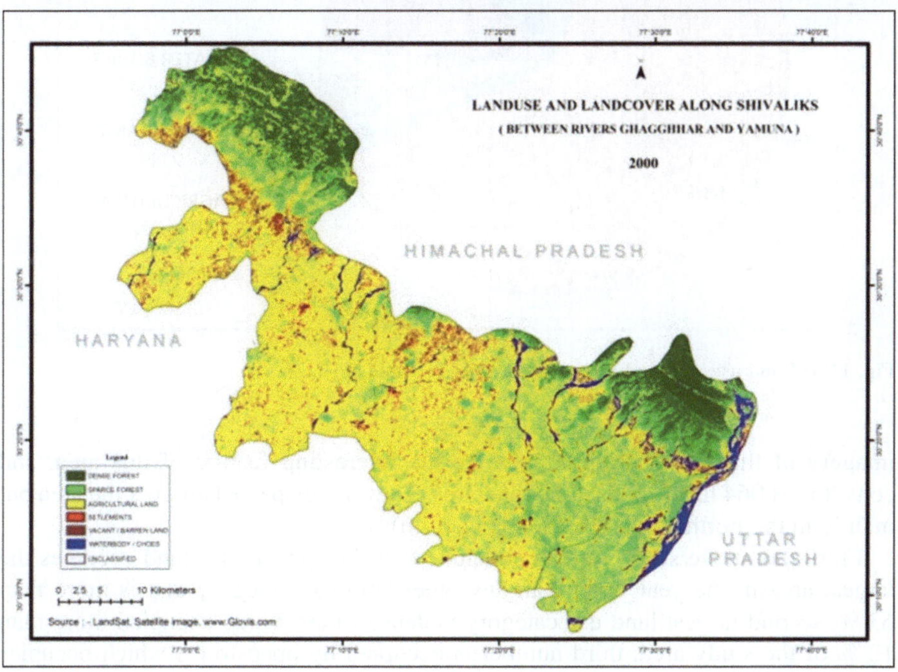

Fig. 11.5 Land use and land cover, 2000. *Source*: IRS, Satellite Data and Landsat Satellite Data, 2000

Table 11.3 Land use and land cover, 2000

S.no	Category	Area in hectares
1	Dense forest	20,862.29
2	Open forest	19,636.56
3	Settlements	7,918.62
4	Water bodies	580.02
5	Barren land	4,719.40
6	Agricultural land	106,108.95
7	Choes	8,064.16
Total		167,890.00

Source: Spatial Analysis with ARC GIS from SOI Toposheet, 2000

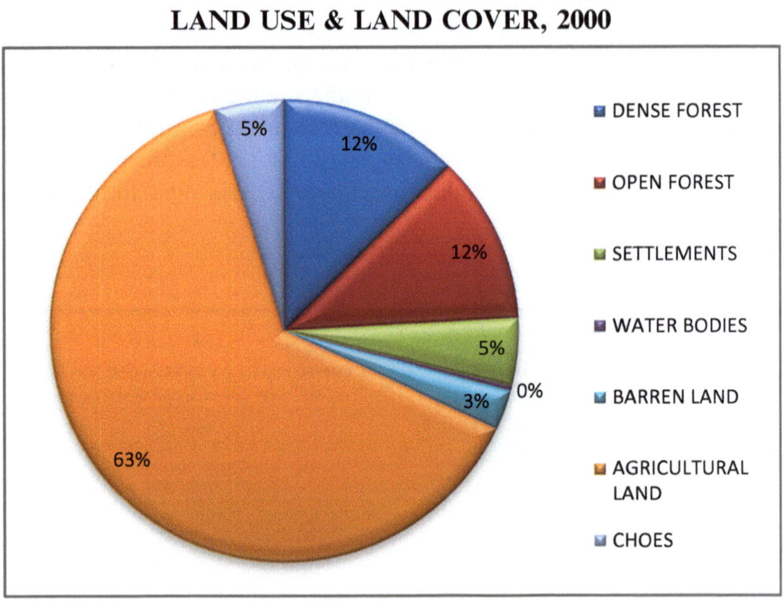

Fig. 11.6 Land use and land cover, 2000. *Source*: Table No 11.3

imagery of this time period. Choes are the interesting feature of this area, and constitute 8,064 ha i.e. 5 % of the total area. They are dispersed all over the area but mainly in the north-south direction (Fig. 11.6).

From the above study we can conclude that the agricultural land occupies the largest area in the year 2000 than any other land use category that is more than 63 %, second largest land use category is dense forest which occupied more than 12 % of the study area, third number is occupied by open forest which occupies about 12 % of the study area. And the least area is occupied by water bodies which is less than 1 %.

11.9 Land Use and Land Cover 2010

In 2010 about 7,393.14 ha of the total land use area was noted down under dense forest. But in 1990 area under the same category was 35,711.62 ha, this means that the total area under dense forests have reduced a greater extent. In this year the dense forests were concentrated in the extreme northern part of the study area. The areas under open forests were found to be increased in the past 20 years. Open forests were distributed in the whole study area. But in large they were found in northern part. About 28,573.31 ha of the total area was found to be under open forest. The total area under agriculture was found to be 113,998.05 ha in the year 2010. A markable change was found in the total area under agriculture in past 20 years, because the total area under this category has increased much more. Some area under this category was also found in the year 2010. But it was also noted that area under this category has increased from 90.10 ha in 1990 to 554.19 ha in the year 2010.

The total area under settlements was found to be 9,869.54 ha. Small sizes as well as large size settlements both were found in the study area. Most of the settlements were found to be concentrated in southern and central part of the study area (Fig. 11.7, Table 11.4).

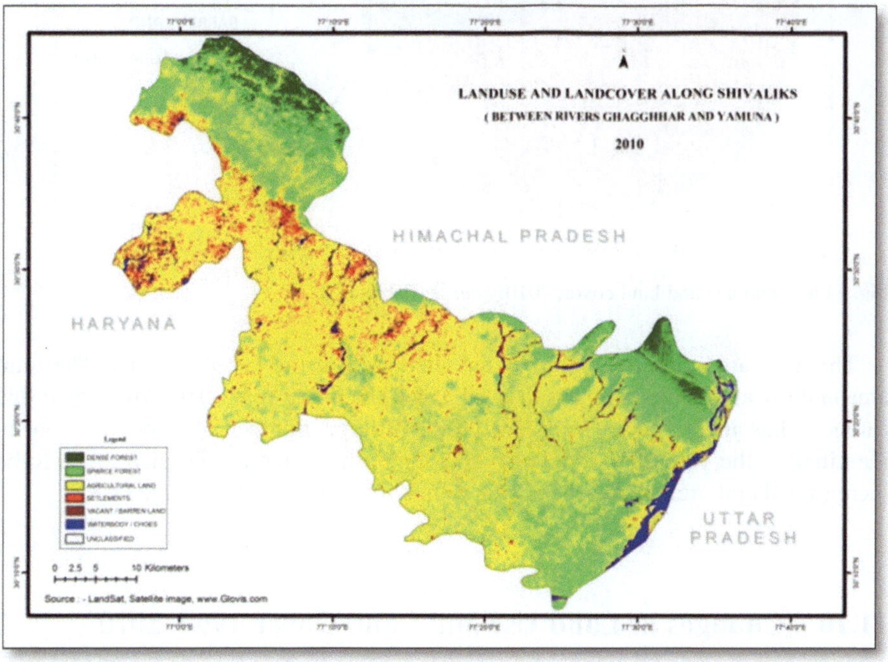

Fig. 11.7 Land use and land cover, 2010. *Source*: IRS, Satellite Data and Landsat Satellite Data, 2010

Table 11.4 Land Use and Land Cover, 2010

S.no	Category	Area in hectares
1	Dense forest	7,393.14
2	Open forest	28,573.31
3	Settlements	9,869.54
4	Water bodies	554.19
5	Barren land	2,361.54
6	Agricultural land	113,998.05
7	Choes	5,140.23
Total		167,890.00

Source: Spatial Analysis with ARC GIS from SOI Toposheet, 2010

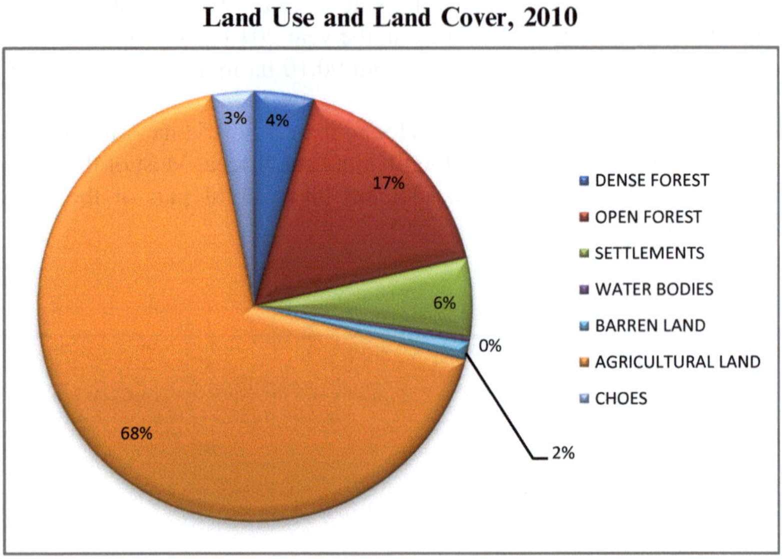

Fig. 11.8 Land use and land cover, 2010. *Source*: Table 11.4

The total area under settlements has increased in the last few years. The total area under water bodies was found to be 554.19 ha in the year 2010. Area under this category has increased in past few years. A large number of choes have been identified in the year 2010. And also area under this category has also found to be increased. Total area under choes was found about 5,140.23 ha (Fig. 11.8).

11.10 Changes in Land Use and Land Cover 1990–2010

Land use and land cover is a changing phenomenon. It is changing since the human started using land as a resource. So, after the years every region is phasing the change into land use and land cover. On the basis of analysis of the area of the year

Table 11.5 Changes in land use and land cover along Shivaliks (Between Rivers Ghagghar and Yamuna) 1990–2010

Land use and land cover classes	1990 (area in hectares)	2010 (area in hectares)	Changes (area in hectares)
Dense forest	35,711.62	7,393.14	−28,313.48
Open forest	1,373.10	28,573.31	27,200.21
Barren land	90.10	2,361.54	2,271.44
Agriculture	105,395.6827	113,998.05	8,602.37
Settlements	1,874.16	9,869.54	7,995.38
Water bodies	17.92	554.19	536.27
Scrubs	12,289.92	[a]	19.4
Choes	11,137.49	5,140.23	5,997.26

Source: Spatial Analysis with ARC GIS, 1990–2010
[a]Category merged with barren land

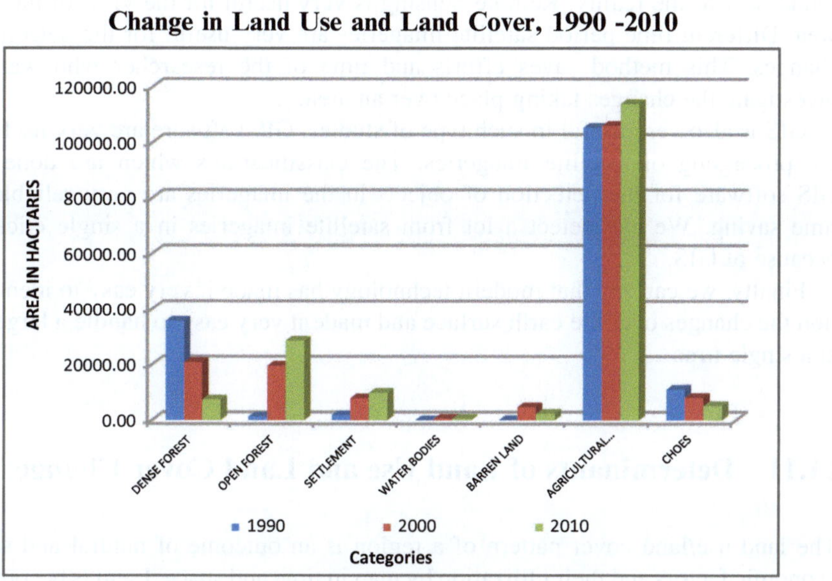

Fig. 11.9 Change in land use and land cover, 1990–2010. *Source*: Table 11.5

1990 and 2010 several changes are also found in land use and land cover of the study region. These changes are found in every land use and land cover class (Table 11.5).

The present study has shown that there remain only few landscapes in the study area those are still in their natural state. Due to anthropogenic activities, the study region is being significantly altered in some manner and man's presence on the area, his use of land has had a profound effect upon the natural environment thus resulting into an observable pattern in the land use/land cover over time. The land use/land cover pattern of a region is an outcome of natural and socio–economic factors and their utilization by man in time and space (Fig. 11.9).

On the basis of discussion in the earlier chapter it can be said that several changes are identified in the study area along the Shivaliks between rivers Ghagghar and Yamuna. These changes are taking place at a rapid rate. The main changes in this area found that land under natural vegetation has decreased to greater extents. Most of the dense vegetation has turned into sparse. The land which was under dense forest in 1990 is found under open forest in 2010. So, this means that the density of natural vegetation is decreasing in this area. Similarly several other changes like increasing area under water bodies, agricultural area, settlements are also found in a noticeable way. An important change in area under these categories is found in the study.

Another objective of the study was that testing of the remote sensing and GIS for the purpose of detection of land use and land cover. This objective is successfully achieved in the study. When the results from remote sensing and GIS were compared with the ground reality they are found quite reliable. The results are found near to the reality. Remote Sensing is very useful for the view of the study area. Different time period satellite imageries are very useful for the detection of changes. This method saves efforts and time of the researcher who wants to investigate the changes taking place over an area.

GIS is also very useful in such type of studies. GIS software are very useful for the processing of satellite imageries. The classifications which are done with GIS software for the detection of objects in the imageries are very reliable and time saving. We can detect a lot from satellite imageries in a single click just because of GIS.

Finally, we can say that modern technology has made it very easy to identification the changes over the earth surface and made it very easy to handle a large area at a single time.

11.11 Determinants of Land Use and Land Cover Change

The land use/land cover pattern of a region is an outcome of natural and socio–economic factors and their utilization by man in time and space. Land is becoming a scarce resource due to immense agricultural and demographic pressure. Land use and land cover of an area is determined by several factors. Determinants of a particular area varies from the determinants of the another area. Similarly the land use and land cover of the study area is determined by several determinants. There are various factors that are responsible for the land use and land cover change alone the Shivaliks between river Ghagghar and Yamuna. Some of the important factors are discussed below:

Encroachment of agricultural land is one of the important factors responsible for the land use and land cover change. Area under agriculture has increased in the past 20 years. The growth of population is one of the major determinants of the land use and land cover change in the present study area. The population of the study area has increased to a greater extent in the past 20 years. The growing population demand for more food grain production, more settlements. Area under natural vegetation has decreased in this area. The main cause of this is the clearance of forests in order to occupy land for agriculture and various other activities and also that people of this area are mainly dependent on fuel wood for cooking purposes. So, they cut natural vegetation for this purpose. They also use grasses and other small plants to feed their domestic animals like buffaloes, cows, goats, sheep, camels etc. The total area under settlements has increased and this is due to the increase in total population of the area. Population has increased to greater extent in this area since last 20 years and due to this increase in population the burden on existing natural resources has also increased. And the basic requirements and needs of the population has also increased so in order to fulfil these requirements the lash green areas were turned to bare. The area under water bodies has increased. Several reservoirs took place in recent year. The area under choes is also decreasing, this is due to people are modifying land for agricultural purposes.

References

Anderson JR, Hardy EE, Roach JT, Witmer RE (1976) A land use and land cover classification system for use with remote sensor data. Geological Survey Professional PaperNo.964, U.S. Government Printing Office, Washington, D.C. p 28

FAO/UNEP (1997) Negotiating a sustainable future for the land. FAO/AGLS, Rome, p 61 [A series of three publications illustrating how to operate participative and sustainable land use planning]

FAO/UNEP (1999) The future of our land: facing the challenge. guidelines for integrated planning for sustainable management of land resources. FAO-AGLS, Rome, p 71 [A series of three publications illustrating how to operate participative and sustainable land use planning]

Khullar DR (2005) Land utilization; India, A comprehensive geography. Kalyani Publisher, Delhi

Mather AS (1986) Land use. Longman Group (FE) Limited, Hong Kong, p 286

Meyer WB (1995) Past and present land use and land cover in the USA Consequences. http://www.gcrio.org/CONSEQUENCES/spring95/Land.html. Accessed 10 Jan 2012

Pandy AC, Nathawat MS (2006) Land use land cover mapping through digital image processing of satellite data. A case study from Panchkula, Ambala and Yamunanagar Districts, Haryana State, India

Petrson JH (2006) Land, work and resources–an introduction to economic geography. Edward Arnold, London

Robinson AH, Morrison JL, Muehrcke PC, Guptill SC (1995) Elements of cartography, 6th edn. Willy, New York

Chapter 12
Land Cover/Biodiversity Change and People Well-Being in Russia

Elena V. Milanova

Abstract The results of land use/cover and biodiversity change and people well-being study in Russia are presented. Agriculture is one of the major driver of land cover and biodiversity change. The main agricultural territory belongs to the steppe zone, where more than 40 % of agricultural production is produced on fertile soils and the major part of the rural population is concentrated. Agriculture restructuring started in country from 1991–1993 have put severe strains on the agricultural production and creates dramatic threats to biodiversity because of plowing mostly of all tillable lands, abandonment or transformation part of them to grazing lands, accelerating of soil erosion, overgrazing and pasture degradation, landscape fragmentation by the fields and infrastructure. Between the global factors that shape the land cover and biodiversity in Russia, climate is the most important one. Climate change scenarios on the base show significant increase of temperature and moderate increase of moistening, that will influence the agriculture production. A complex present-day landscape methodology is used to study land cover and biodiversity changes under agricultural impact in Russia. Due to the ongoing agriculture restructuring reforms in Russia local communities have to manage an outdated and inefficient economic and social infrastructure that previously was maintained by federal budget money and the living standards of people are still lower than in the most developed countries. The ecologically sound land use practice and biodiversity conservation should help to reconcile social needs with the requirements of the environment and to avoid the unfavorable consequences of land degradation.

Keywords Agriculture • Biodiversity • Land use/cover change • Landscape methodology • People well-being • Russia

---

E.V. Milanova (✉)
Moscow State University, Faculty of Geography/Fund for Sustainable Development, Moscow, Russia
e-mail: emilanova@fund-sd.ru

M. Singh et al. (eds.), *Climate Change and Biodiversity: Proceedings of IGU Rohtak Conference, Vol. 1*, Advances in Geographical and Environmental Sciences,
DOI 10.1007/978-4-431-54838-6_12, © Springer Japan 2014

12.1 Introduction

The results of land use/cover and biodiversity change and people well-being study in Russia are presented. The research was carried out at the Faculty of Geography of Moscow State University on the base of the combination of remote sensing and in-field data of different spatial and temporal resolution. In Russia the major driver of land use/land cover change is agriculture. Twenty years ago modernization of Russian agriculture has been started. In some regions agricultural lands were dramatically impacted by changed management practices, resulted in accelerating erosion and reducing biodiversity. Climate is the most important global factor that shape agriculture and biodiversity. A complex present—day landscape methodology is used to study the land cover/ biodiversity change under agricultural impact in Russia as well as to explain relationships between land use practice and land cover patterns. The analysis of the present status and trends of evolution of natural and anthropogenic landscapes permits the study of long-term and short-term land use/cover dynamics. Feasibility study had been undertaken for scale-dependent landscapes applications and study of land cover dynamics under ongoing changes in Russia. The healthy environment cannot be made better in poverty, and poverty cannot be eliminated without sustainable healthy environment, that's why the sustainable development is closely connected with the people well-being. The ecologically sound land use practice and biodiversity conservation should help to reconcile social needs with the requirements of the environment and to avoid the unfavorable consequences of land degradation.

12.2 Land Use/Cover and Biodiversity Change Main Drivers

12.2.1 Agriculture: One of the Major Drivers of Land Cover and Biodiversity Change in Russia

Agricultural lands occupy more than 220 million ha (13 % of the total country territory). The structure of agricultural lands exhibits significant geographical diversity. The most of Russia is located in the forest and tundra vegetation zones, and is considered as a high-risk zone for agriculture. In the northern and north-eastern part of the country, the agriculture (primarily growing grasses and cereals for fodder) is limited with permafrost and insufficient growing degree days. In the south agriculture is limited by the arid and semi-arid climate conditions.

The main agricultural territory belongs to the steppe zone and represents the wide (1.5 thousand km) latitudinal belt to the south of 55°33′N in-between broad-leaf forests and semi-desert zones, going further to the south in Kazakhstan and Mongolia. Along the north-south transect, the steppe zone changes from the forest-steppe in the north to steppe subzone in the south. The steppe is the most important for food production vegetation zone in Russia. Fertile chernozem soils and moderate climate of the steppe region ensure that more than 40 % of the total Russia's harvest is produced here. The region provides employment for 30 % of the total rural population of the country. In West Siberia, steppe zone is much narrower, with forest steppe strips. Further east, in Central and East Siberia, the steppes are mostly localized in depressions and on the low parts of the mountain slopes.

The steppes are characterized by high level of biodiversity which is considerably (4–5 times) decreasing from west to east and from north to south—more high in European steppes than in RFE (Dauria) steppes (from 1,400 to 400 plant species).

Until the eighteenth century, agriculture in Russia was limited to the forest zone. Agricultural practice in the forest zone demanded extraordinary efforts from the peasants providing very limited and highly variable yields due to short growing seasons and the extremely long (over 200 days) cold period—that's why later the Russian peasants started to relocate into the rich chernozem soil lands of the forest-steppe and steppe zones.

The reform in Russian agriculture started in 1991–1993 with changing the property rights. Up until that time, the land had been under federal ownership; now, the farmers received a partial ownership, however little change in land use followed: similar to the Soviet practice, Russia's agriculture is still dominated by the former collective farms (up to 62 % – 138 million ha), relatively small area of the land had been granted to the individual farmers. In terms of operational management (current decision–making) and their direct usage the picture looks somewhat different—the bulk of agricultural lands is managed by the elected bodies of small-scale groups (municipal entities of rural settlements in association of land share owners-stock societies, agricultural production cooperatives, etc.). Only 20 % of these lands are managed more or less independently by private land users (owners or lessees).

The agricultural production on plots of land belonging to individual farmers is considerably more intense than that on the collective farms' lands. Even though the total area of agricultural land in individual ownership is only 13–24 % (depending on the crop), its contribution to agricultural production is disproportional higher. Frequently, the property rights for farmland are being acquired (usually through leasing) by large investors, such as the oil companies or banks. During the denationalization process, the lands (usually the grasslands and ranges) were also allotted to municipal entities and are owned by rural administrations.

The study of strong human impact on steppes landscapes is of high importance because as it was mentioned above more than 40 % of agricultural production of the country is produced in this zone with fertile soils, where the major part of the rural population is concentrated. The transformations of steppe landscapes are related with their conversion to cultivation (about 90 %), large industrial installations (hydropower stations) and urbanization. The changes of land cover were caused also by political and economic transition conditions: changes in ownership, stagnation in agriculture, development of transport infrastructure. The legislation limits the use of these lands besides to agricultural production regardless of the form of property to such lands. At the same time non-use of agricultural lands is officially forbidden.

12.2.2 Agriculture and Biodiversity

Agricultural intensification creates new threats to biodiversity. That includes plowing of all tillable lands, which increases the rate of soil erosion, overgrazing, which increases pasture degradation, landscape fragmentation by the fields and infrastructure, etc. Degradation affects more than 70 % of land under agriculture.

On the other hand, the role of agricultural lands in preserving steppe biodiversity is of particular importance as close to ¼ of the entire steppe zone has already been converted into the semi-natural agricultural landscapes: meadows, pastures, hay fields, and fallows. Many species live predominantly or exclusively on these agricultural lands. That includes more than 6,000 species of plants, about 100 species of mammals, 150–180 species of birds and thousands species of insects and other invertebrates found in the steppe grasslands which become endangered if the current practice of agricultural management is changed.

Although the monetary value of biodiversity is rarely taken into account, maintaining biodiversity is important for both agriculture and well-being of local communities. Semi-natural and natural fodder fields area in Russia is more than 70 million ha (including 26 million ha in steppe and forest-steppe), or 32 % of all fodders, and serve to produce hay and other phytomass. Another example of the economic importance of biodiversity is bee keeping (with wild plants growing in steppe lands converted to agriculture serving as melliferous herbs), medicinal plants, edible plants, and mushrooms. Semi-natural landscapes are also important for ensuring preservation of plowed field fertility and self-recovering of territories.

12.2.3 Climate Change Influence on Land Use in Russia

Between the natural factors that shape the land cover and biodiversity in Russia, climate is the most important one. The five different Global Circulation Models (GCMs) were used for three pre-set time periods: 2020s, 2050s, and 2080s (Intergovernmental Panel on Climate Change 2007). All climate change scenarios show significant increase of temperature and moderate increase of precipitation. Since the effect of global warming is even more profound at high latitudes, the northern countries, like Russia, can reasonably expect even higher temperature rise. There is no doubt that such high impact will significantly affect the agricultural sector, but at what rate, or even negatively or positively, is still under research. The reason for that is that the effect of global warming is at least twofold. On one hand, higher temperatures will increase the length of vegetation period and reduce the risks connected with spring and winter frosts. On the other, increasing temperature with decreasing precipitation will lead to higher probability of summer droughts.

Taking into account that the steppe zone of Russia is the principal agricultural producer of grain for domestic and export requirements, any research of climate change impacts should include the changes in the risk of droughts.

In 2020s the annual temperature is estimated to increase by 1.8 °C for the entire Russia and by 1.6 °C (compared to 1961–1990 baseline temperature) for the main steppe agricultural zone; in 2050s, the annual temperature is simulated to increase by 3.4 °C for the entire Russia and by 3.1 °C for main agricultural zone of the country. The annual temperature in 2080s is expected to increase by 4.7 °C for the agricultural regions on average (5.2 °C for the entire Russia).

Annual precipitation in 2020s increase by 32 mm on average, with small increase or decrease in the difference between potential evapotranspiration and precipitation, a value that provides information on moisture deficit. For the areas of high and considerable development of agriculture, the precipitation increase is also smaller. In 2050s -precipitation increase by 14–65 mm, i.e. by 3–13%, compared to the baseline climate. The increasing water deficit leads to aridization, which affects the most south regions of country and threatens the conventional farming. The precipitation in the agricultural regions is projected in 2080s to increase by 44–70 mm; the further decrease of water availability for the rain-fed agriculture is expected in south regions.

12.3 Methods and Results of Land Cover Classification

12.3.1 Present-Day Landscape Approach

The landscape approach to the environment as a combination of hierarchically subordinated geosystems (present-day landscapes) was used for realistic understanding and study of land use/land cover dynamics and changes in steppe regions. It provides a basis for the perception of the world as a system of interrelated territorial samples with different combinations and close interaction of natural and socio-economic components (Milanova et al. 1993).

The landscape analysis represents a holistic view on the man-made and natural surroundings and attempts to bridge the gap between natural and human subsystems. Within the framework of present-day landscapes mapping it will be possible and relevant to investigate how regulatory tools at various political or administrative levels addressing different parts of the systems will influence the environment and how sensible different parts of the landscape are to human driving forces as to regulatory instruments (local, regional, national, or international). The reason for the complexity of the systems needs further investigations to ensure, that more simple analytical approaches will not suffice. A likely outcome is that the systems are complex due to the fact that social, economical, technical and ecological processes operate on a wide range of spatial and temporal scale.

12.3.2 Scale-Dependent Land Cover Applications Through Remote Sensing Technique

At country level investigation at this level was implemented for the the whole territory of the former Soviet Union (40°−80°N, 20°−180°E). The most sufficient source of NOAA-AVHRR NDVI images for continental-scale feasibility study was the Global Ecosystems Database (GED), Version 1.0 (on CD-ROM) by EPA Global Climate Research Program, NOAA/NGDC Global Change Database Program (Kineman and Ohrenschall 1992). Experiments with present-day landscape mapping and classification done on the basis of remote sensing at this macroregional level have shown the following features of steppe natural zonal structure and its anthropogenic transformation (Milanova et al. 2005).

According to analysis of vegetation biomass seasonal dynamics based on 10 km resolution data set, zonal stratification of landscapes very often differs from that which is traditionally drown on vegetation and landscape maps. Preliminary experiments in this field show that human activities have changed the pattern of natural zones in many regions of Russia. As far as vegetation activity is concerned landscapes of sub humid steppe here greatly differs from dryer steppes of the Volga region and West Siberia, although traditionally they are shown on vegetation and land cover maps as the same natural zone without further subdivision.

For regional level the case study area in Volga steppe region. The map showing the state of agricultural lands in Samara oblast was compiled using space image which was received from the Russian «Resurs-0» module. The results of land classes identification by means of their spectral curves analysis allowed determining areas including agricultural lands characterized by different status of vegetation cover: the lands with different crop covers are shown: lands with satisfactory crop cover of soil is more than 50–60 %, moderately suffered lands with crop cover of soil is 20–50 % and strongly or catastrophically suffered lands with crop cover is less than 10–20 %.

The Middle Volga Region is characterized by variability of climatic conditions and due to this—by considerable fluctuations in crop yields. Recurrent draughts caused severe failures of crops. One of the most severe draught occurred in 1891 and failure of crops caused mass starvation in a number of districts in the Volga Region. About one third of arable lands lacked in crop-shoots. The effects of failure of crops in 1921 were increased by political factors and resulted in millions of victims in the Volga Region. In 1946, the whole region suffered from failure of crops caused by drought. The last severe drought in the region occurred in 1998. It caused a catastrophic failure of crops. About one third of arable lands lacked in crop-shoots.

12.4 Sustainable Environment and People Well-Being

The healthy environment cannot be made better in poverty, and poverty cannot be eliminated without sustainable healthy environment, that's why the sustainable development is closely connected with the people well-being. Due to the ongoing municipal reforms in Russia in economy sphere, especially in agriculture, the local communities faced and have to manage an extremely outdated and inefficient economic and social infrastructure that previously was maintained by federal budget money. During the last decade all over the world the traditionally dominant ideas of material product-based well-being started to be changed by idea of immaterial "access-based" well-being (availability of access to immaterial goods: to services, information, education, to places of pleasure). The similar changes are happening in Russia during the period of economy restructuring and market economy development. The ecologically sounds land use practice and biodiversity conservation should help to reconcile social needs with the requirements of the environment and to avoid the unfavorable consequences of land degradation.

Russia is the country with rather specific attitude to notion of common goods. This notion is translated in Russian as consuming goods. The term of "so-called common" (in proper sense) goods existed in Russia during the long socialist time period of history (70 years) and stands for "goods" belong for everybody (in reality mostly to state) and nobody in particular. These "common" goods included lands, environmental and other resources, equipments of collective and soviet enterprises. As long as they remained "common", they cannot be mostly marketable products. During the last decade all over the world the traditionally dominant ideas of material product-based well-being started to be changed by idea of immaterial "access-based" well-being (availability of access to immaterial goods: to services, information, education, to places of pleasure), what is more coincident with ideas of freedom (Concerted Development of social cohesion indicators 2005).

The similar changes are happening in Russia during the period of economy restructuring and market economy development. These changes are linked with increasing of country openness and relationships with other world, environmental degradation, and weak local self-governance in regions. For Russian people during mostly of the whole country history immaterial (so-called spiritual) values were always very important (traditional culture of different nationalities, huge country lands and spaces, environmental richness and spiritual wealth).

Anyway in Russia the living standards of people are still lower than in European and other developed countries, and differentiation in wealth conditions of different parts of population is quite high and quickly strengthening. That's why material goods still continue to be also rather important for people well-being.

12.4.1 Projects on Community Sustainable Development and Their Influence on People Well-Being

In cooperation with Russian NGO "Fund for Sustainable Development" (FSD) the community development projects were implementing on municipal level in different part of Russia. The community success formula is defined by two parameters: (1) community improvements (optimized energy consumption; improved life quality; better environment) and (2) enhancing community processes (public, social and economic) including expanding multi-stakeholder cooperation; community involvement in decision-making; increasing transparency of local policy.

The results/goods of ecologically sounds projects (on biodiversity and nature conservation, agriculture modernization, energy efficiency, ecotourism development and others) could be conditionally /roughly differentiated into material and immaterial ones: (1) material goods (new ecologically sounds technologies and installations, local healthy agricultural products, constructions for ecotourism development) and (2) immaterial goods (access to social services, information and education, access to natural values of biodiversity and landscapes under conservation, people traditional culture).

The following main indicators for evaluating of projects' influence on community success and people well-being were elaborated:

- environment improvement (waste/pollution reduction and environmental health improvement, amelioration of agricultural lands, development of recreation areas/parks as well as ecotourism and environmentally responsible business).
- economic improvement in agriculture, energy saving and housing sectors (introduction of new technology in agricultural and other economy sectors reinvestment of energy savings into solving of community problems);
- social improvement (growth of people well-being, healthy environment and agricultural production, reduction of energy rates for people, lower unemployment level);
- public & legislation (modernization of land use practice, biodiversity and natural resources conservation, public participation in adoption of municipal laws and decisions);
- health (health improvement in the community (better and more comfortable conditions in child care institutions, quality of drinking water improvement);
- education (people knowledge strengthening on land and other natural resources conservation, altered people behavior models).

12.5 Conclusion

Land Cover and biodiversity change researches have to address relationships of global challenges and local consequences/solutions, which have to coincide with each other. Local solutions have to take into consideration local nature and economic conditions to elaborate sustainable land use policy. Agriculture is one of the major drivers of land cover and biodiversity change in country. Agriculture restructuring have put severe strains on the agricultural production and creates dramatic threats to biodiversity. The present-day methodology of land cover classification basing on remote sensing data and analysis of all currently existing data (local and regional maps, satellite imagery, detailed field observations) allowed to produce up-to-date land characteristic database and to study land cover/land use structure and trends of their changes. Considering and measuring well-being indicators there were taken into account not only such people well-being criteria as economic, environmental and social issues, but also personal well-being perception (or live satisfaction). The results of research has direct relevance to land management and nature protection in agricultural steppe belt in Russia, that should help to reconcile social needs with the environment requirements and to avoid the unfavorable consequences of lands and biodiversity degradation.

References

Concerted Development of Social Cohesion Indicators (2005) Council of Europe Publishing, Strasbourg, p 235, ISBN 920871-5742-1

Intergovernmental Panel on Climate Change (2007) Climate change 2007: the physical science basis. Contribution of working group I to the fourth assessment report of the intergovernmental panel on climate change, Cambridge University Press, Cambridge, United Kingdom and New York

Kineman JJ, Ohrenschall MA (1992) Global Ecosystems Database. Version 1.0 (on CD-ROM). Documentation manual. EPA global climate research program, NOAA/NGDC global change database program

Milanova EV, Kushlin AV, Middleton N (eds) (1993) World map of present-day landscapes. Soyuzkarta, Moscow

Milanova E et al (2005) Landscape approach for land-use/land-cover classification and mapping at different scales. In: Elena M, Yukio H, Ivan B (eds) Understanding land-use and land-cover change in global and regional context. Enfield and Plymouth, Science Publishers Inc, pp 233–247

[illegible]

Chapter 13
Public Perception on Endangerment of Hornbills: A Field Study on Upper Siang District of Arunachal Pradesh

Miyo Tayeng, Philip Mody, and S.K. Patnaik

Abstract Arunachal Pradesh enjoys the unique status of being the state with the wildest spectrum of wildlife species. It may be attributed to its peculiar biogeography location, altitudinal variation and high rainfall. Altitudinal differences gave rise to different climatic regimes and soil structure, which in turn has determined the spectacular vegetations of the state. Infact, the land manifests a phenomenal range of rich biological diversity. Perhaps avian fauna best represent this diversity of which *Bucerotidae*- a hornbill is of utmost importance. The state has five important species of hornbills out of nine being found all over in India. Fascinatingly, a hornbill is the state bird of Arunachal Pradesh because of its widespread distribution in the state and deep social significance among most of the people rather than as a most chosen food. However, hornbill has been declared endangered species of the state as its counts keeps on dwindling down due to either rapid destruction of its habitat or excessive hunting in the recent times. Thus, necessitated its immediate conservation.

The present paper makes an attempt to give an overview of various species of hornbills being found and its distribution across the Arunachal Pradesh. Further, it makes detail discussion on public's perceptions of Upper Siang district of Arunachal Pradesh on endangerment of hornbills.

Keywords Arunachal Pradesh • Biodiversity • Hornbill and Upper Siang

M. Tayeng • P. Mody (✉) • S.K. Patnaik
Rajiv Gandhi University, Itanagar, India
e-mail: philipmodi@rediffmail.com

M. Singh et al. (eds.), *Climate Change and Biodiversity: Proceedings of IGU Rohtak Conference, Vol. 1*, Advances in Geographical and Environmental Sciences,
DOI 10.1007/978-4-431-54838-6_13, © Springer Japan 2014

13.1 Background of Discussion

Arunachal Pradesh enjoys the unique status of being the state with the wildest spectrum of wildlife species, which may be attributed to its peculiar biogeography location, altitudinal variation and high rainfall. A unique feature of the state is that it still has fair chunk of area, within and outside protected area network that is in undisturbed state. Indeed, altitudinal differences gave rise to different climatic regimes and soil structure, which in turn determine the vegetation and faunal condition for growing wide ranges of flora. It offer good shelter and ideal niche to varieties of fauna and thus, the land manifests a phenomenal range of biological diversity of which the avian fauna best represent this diversity. The state has amazingly rich avifauna with over 760 bird species reported from the area. Most of the species are resident of the state and undergo altitudinal migration seasonally, though some winter migrants from the Palearctic region like waterfowl, waders, wagtails etc., are common in wetland areas. The state has distinction of having many rare species of the birds reported nowhere else in India like Ward's Trogon, Spotted Long-tailed Wren Babbler, Austen's Babbler, *Mishmi* Wren (which occurs nowhere else in the world) etc. Many endangered species of birds like Hornbills, White-winged Wood Duck, Black-necked Crane, Bengal Florican, Swamp Patridge etc. have been reported from this state. Of these endangered avian species, *Bucerotidae* commonly known as hornbill is in receiving ends of extreme threat to extinctions. Its counts are rapidly on decrease over the years in Arunachal Pradesh due to myriads of causative mechanism. The beak of hornbill is used as head decor or gear as a part of traditional attire by major chunk of tribal populace of this state. Moreover, it is hunted for food and has indelibly entered into taste buds of most of the local people. Above all, increasing numbers of licensed guns at the hand of local people has aggravated situation in the state.

Hornbill is in receipt of threats of being lost due to various telltale reasons. Vanishing habitats of hornbill is frantic in the state. Conversion of forest land into farm lands, increasing practices of shifting cultivation and uncontrolled logging are some of them. To great dismay, it is as good as not seen bird species in central and eastern parts the state owing to excessive hunting and losts of its habitat. The situation is better in *Namdapha* tiger reserve and foothill forests of *Pakhui* national park and in *Khellong* forest division. Local taboos on hunting during the breeding season (March to July/August) by *Nishis* in the *Seijosa* area have probably resulted in conserving this rare species.

13.2 Objectives

This paper endeavor to:

1. Provide an overview of hornbill types and its distribution in Arunachal Pradesh.
2. To analyze public perception on endangerment of hornbills in Upper Siang District of Arunachal Pradesh.

13.3 Universe of Study

The study area stretch over one district namely-Upper Siang of Arunachal Pradesh. The present study has been conducted in 5 villages namely, *Mariyang, Damroh, Millang, Yingkiong and, Geku* respectively. The facts and figures pertaining to hornbill over the last 2 years i.e., from 2011–2012 to 2012–2013 have been taken into consideration for the present study.

13.4 Research Methodology

During the present study an effort has been made to make the study empirical based on survey and statistical methods. The work is purely based on both primary and secondary data.

For the collection of primary data, a field survey was conducted over 5 villages and 6 respondents from each village were interviewed personally with the help of well-designed schedule. Altogether, 30 respondents comprises of public leaders, student, animal hunters, general public, Govt. servants and business men have been selected at random basis and interviewed to ascertain their perception on hornbill-its conservation and endangerment. As regard to secondary data, reports and publication of State Forest Research Institute, Govt. of Arunachal Pradesh has been used extensively for the present study. In addition to this, various published books, research papers and articles on hornbills have been referred to for better understanding of the research problem. However, internet has been one of the most used sources of secondary data for the present study. Primary and secondary data has been analyzed, interpreted and summarized with the help of various types of statistical tools like percentage, mean etc. Audio-visual tools viz. tape recorder, and digital camera have also been used to facilitate the collection and recording of data during the field study. For analysis of field data, SPSS and MiniTab have been used for the present study.

13.5 Hornbills in Arunachal Pradesh- Types & Distribution

India is a home to 9 off 49 hornbill species of the world. However, in Arunachal Pradesh 5 species are found namely- Rufous Necked Hornbill, Wreathed Hornbill, Oriental Pied Hornbill and, Great Hornbill respectively. The Table 13.1 shows various hornbill species found in the state with its biographic details and information on habitat.

Table 13.1 Hornbill species of Arunachal Pradesh

Hornbill species	Body mass	Habitat
Austen's Brown hornbill	Male: 933 g. Female:755 g (Kemp 1991)	Dense evergreen forest from plains to 900 m
Rufous Necked Hornbill	Male 2,500 g (Kemp 1991)	Hill evergreen forest from 500 m up to 2,100 m
Wreathed Hornbill	Male 2,515 g, female 1,950 g (Kemp 1991)	Lowland foothill semi-evergreen and evergreen forest, but also up to 1,800 m
Oriental Pied Hornbill	Male 738 g. female 624 g (Kemp 1991)	Forest edge, open moist deciduous and evergreen forest, reverie forest. Secondary logged forest s and even gardens and agricultural fields.
Great Hornbill	Male 3,007 g, female 2,211 g (Kemp 1991)	Primary evergreen forests and moist deciduous forest, mainly on lowland plains but can extend up to 2,000 m. Also seen in selectively logged forests and plantation close to larger forested tracts.

Source: State Forest Research Institute, Arunachal Pradesh: Itanagar

Austen's Brown hornbill or Brown hornbill (*Anorrhinus austeni*) (also known as the White-throat brown hornbill, formerly *Ptilolaemus tickelli*) are found in the countries like- India, Myanmar, Thailand, Laos, Vietnam and South India. In India, restricted to eastern Arunachal Pradesh and Assam. Reported from *Namdhapa* Tiger Reserve (TR), evidence seen in *Jairampur* Forest Divison in *Changlang* District, and in lower areas near *Deomali* and *Nakfan* in *Tirap* District. Also sighted from areas in Upper Assam in *Joypur* Reserve forest. *Tinsukia* District and in *Cachar* Hills and from several other reserve forest in upper Assam. Distribution in other areas of north-east India is inadequately known, though it may occur or have occurred in Nagaland and Manipur. Oriental Pied Hornbill (*Anthracoceros albirostris*) also known as Indian Pied hornbill earlier are found in South Nepal, South Bhutan, North Bangladesh, northern and north-east India, Myanmar, Mergui archipelago, South China, Vietnam, Laos, Cambodia and parts of Thailand, North-East Peninsular Malaysia. Rufous-necked hornbill (*Aceros nipalenis*) are found in Nepal (believed extinct), north-east India, Bhutan, east Myanmar, north & west Thailand, South China, north Laos and north Vietnam. Unconfirmed in Cambodia. Wreathed hornbill (Aceros undulates) also known as Bar Puched Wreathed hornbill are found in North-East India, South Bhutan, Myanmar, Thailand, Cambodia, Vietnam, Laos, Peninsular Malaysia, Indonesia on Sumatra and adjacent Islands, Java, Bali, Kalimantan, Sarawak, Sabha, Brunei, and other smaller islands. Great hornbill (Bucerous bicornis) also known earlier as Great Pied or the Great Indian hornbill are found in Western Ghats, and from Himalayan foothills in Uttaranchal, to south Nepal, Bhutan and north-east India. Myanmar and some islands in the Mergui archipelago, South China, Vietnam, Laos, Cambodia, Thailand peninsular Malaysia and several adjacent island, an isolated population in Indonesia on Sumatra.

13.6 Public Perception on its Endangerment: Field Experience

It is observed from Table 13.2 that public of Upper Siang District has received no financial supports from state Govt. towards conservation of hornbill. Moreover, 27 off 30 respondents which constitute 90 % of sample respondents hold views that they didn't yet received any technical supports from the Govt. On positive note, 2 persons which constitute only 6.7 % have perceived that state Govt. promotes tourism on hornbill. Further, one person that represent only 3.3 % of total samples of present study have opined that other supports provided by state Govt. are simply alright.

As regards to public awareness is concerned, public of the study area are found to have least aware of other supports like- propagandas, publicity, workshops etc. on the conservation of hornbill. 43.3 % of the total respondents are found to have not aware of other supports provided by state Govt. apart from financial, technical supports etc.

As evident from Table 13.3, 53.3 % of respondents have mildly agreed upon endangerment of hornbill in the study area. 36.7 of the total respondents have strongly agreed that hornbill is diminishing in Upper Siang District.

Sixty percent and 30 % of the total respondents simply disagree and strongly disagree on adequacy of various facilities provided by state Govt. for conservation of hornbill.

In addition, 53.3 %, 36.7 % and 10 % of the sample respondents strongly agree; simply agree and; neither agrees nor disagrees upon decreasing trend of hornbill counts. Further, 50 % and 43.3 % of the respondents neither agrees nor disagrees and simply agreed on uses of artificial hornbill beak for social reasons.

Moreover, field study revealed that 76.7 % of total respondents neither agrees nor disagrees upon avoidance of hornbill hunting. However, 23.3 % of respondents mildly agreed that hunting of this endangered species be abandoned.

Table 13.2 Public perception on State Govt. support on Hornbill

		No facility	Poor	OK	Good	Don't know
Financial support	Frequency	30	–	–	–	–
	Percent	100	–	–	–	–
Technical support	Frequency	27	2	–	–	1
	Percent	90	6.7	–	–	3.3
Conservation support	Frequency	18	9	–	–	3
	Percent	60	30	–	–	10
Tourism promotion	Frequency	9	14	–	2	5
	Percent	30	46.7	–	6.7	16.7
Other supports	Frequency	13	3	1	–	13
	Percent	43.3	10	3.3	–	43.3

Source: Field study

Table 13.3 Public perception on Hornbill

		Strongly disagree	Disagree	Neither agree nor disagree	Agree	Strongly agree
Hornbill is an endangered species	Frequency	–	–	3	16	11
	Percent	–	–	10	53.3	36.7
State Govt. provide adequate facilities for conservation of hornbill in our area	Frequency	9	18	2	1	–
	Percent	30	60	6.7	3.3	–
Numbers of hornbill keeps on decreasing in our region	Frequency	–	–	3	11	16
	Percent	–	–	10	36.7	53.3
We should conserve hornbill using traditional knowledge and skills	Frequency	–	–	7	19	4
	Percent	–	–	23.3	63.3	13.3
We should use artificial beak of hornbill for our social reason	Frequency			15	13	2
	Percent			50	43.3	6.7
We should avoid hunting hornbill				23	7	
				76.7	23.3	

Source: Field data

During the field study it is revealed that hornbill is traditionally conserved through restricting other community from hunting into community land in the study area. *Kebang* (village council) of the study area is also found active in restricting hunting of hornbill in many villages. *Kebang* restrict hunting of this species especially in community land that helps in conservation of hornbills. Imposition of fines by village council when somebody engage in hunting of hornbill either in community land or any other forest area beyond festival season is found to have great bearing in conserving this species. Moreover, public of the study area are found to have given up hunting of hornbill during fruit bearing seasons of *Tapil* (fruits eaten both by local community and hornbill). During field study, it is also unveiled that growing numbers of licensed guns among people is responsible for orgy of hornbill hunting.

13.7 Findings

1. General public of Upper Siang District has received no financial supports from state Govt. towards conservation of hornbill till date.
2. Ninety percent of sample respondents didn't yet received any technical supports from the state Govt. on conservation of hornbill.
3. Only 6.7 % have perceived that state Govt. promotes tourism on hornbill in the study area.
4. 53.3 % of respondents have mildly agreed upon endangerment of hornbill in the study. Moreover, 36.7 of the total respondents have strongly agreed that hornbill is diminishing in Upper Siang District.

5. Sixty percent and 30 % of the total respondents simply disagree and strongly disagree on adequacy of various facilities provided by state Govt. for conservation of hornbill.
6. 53.3 percent, 36.7 % and 10 % of the sample respondents strongly agree; simply agree and; neither agrees nor disagrees upon decreasing counts of hornbills in the study area.
7. Fifty percent and 43.3 % of the respondents neither agrees nor disagrees and simply agreed on uses of artificial hornbill beak made of fiber for social reasons.
8. Moreover, field study revealed that 76.7 % of total respondents neither agrees nor disagrees upon avoidance of hornbill hunting. However, 23.3 % of respondents mildly agreed that hunting of this endangered species be abandoned.
9. Further, community of the study area is found to have been conserving hornbill using their traditional knowledge which is administered through village council called *Kebang*. This village council restricts and imposes fines on those who engage in hornbill hunting into community land and beyond festive seasons etc.

13.8 Concluding Remarks

Due to peculiar biogeography location, altitudinal variation and high rainfall, Arunachal Pradesh has the wildest spectrum of wildlife species in India. Avian fauna best represent this rich pool of wildlife of which *Bucerotidae*- a hornbill is of current importance. As hornbill is declared endangered species of the Arunachal Pradesh as its counts keeps on dwindling down due to either rapid destruction of its habitat or excessive hunting in the recent times. To great dismay, State Govt. is yet to provide financial supports, technical supports, conservation supports etc. to the people of Upper Siang region. People of the study area hold strong conviction that it is an endangered species. It is rarely found in the study area. It occasionally visits the study area especially during fruit season of *Tapil*.

Nevertheless, local community of Upper Siang has evolved aged-old mechanism to conserve hornbill which is basically administered through village council-*Kebang*. It is also felt that state Govt. should de-license guns to curb hunting of hornbill in the study area so that glory of Upper Siang in particular and Arunachal Pradesh in general in terms of rich bio-diversity is retained forever.

References

Annual Report, State Forest Research Institute, Itanagar, Arunachal Pradesh, 2012

Kemp A (1991) Forshaw J (ed) Encyclopaedia of animals: birds. Merehurst Press, London, pp 149–151

Chapter 14
Potential of Earthworms in Bioconversion of Organic Solid Waste

Jyoti, Vineeta Shukla, and Seema Rani

Abstract The generation of waste materials is increasing proportionately with the growth of human population and increasing pace of industrialisation. Millions of tonne of solid waste generated from the modern society are ending up in the landfills every day, creating extraordinary economic and environmental problems for the local government to manage and monitor them for environmental safety. The methodology of solid waste management has shifted from conventional disposal strategies such as incineration, landfill etc. to conversion of waste into value added products during recent years. The usage of solid waste by recycling can supply nutrients to vegetative plants and also improve soil physical conditions and its fertility. Earthworm with their marvellous capability of ingestion, digestion and excretion are nature's most useful converters of wastes. Earthworm participation enhances natural biodegradation and decomposition of solid waste from 60 to 80 % thus significantly reducing the composting time by several weeks. To reduce the cost of disposal of solid waste and best utilization, it was planned to convert the solid waste into a valuable vermicompost. Consumption of organic waste earthworm culture is an ecologically safe and economically viable process to get beneficial products. While they devour our organic waste, thus decreasing our disposal problems, they are also and concurrently manufacturing two new products-earthworm biomass and vermicompost.

A laboratory experiment was carried out for proper management of solid waste through the action of indigenous earthworm, *Metaphire posthuma* of mixtures containing solid waste and cow dung. The action of worms accelerated the decomposition of wastes. Analysis of soil bed and waste from experimental container after 15 days interval for physical and bio chemical activities revealed that worm is capable of recycling of solid waste into useful nutrients. During this process organic matter, pH and C:N ratio revealed negative trend, however total nitrogen, phosphorous and potassium content expressed positive trend of increment with

Jyoti (✉) • V. Shukla • S. Rani
Department of Zoology, M. D. University, Rohtak, India
e-mail: dalaljyoti10@gmail.com

M. Singh et al. (eds.), *Climate Change and Biodiversity: Proceedings of IGU Rohtak Conference, Vol. 1*, Advances in Geographical and Environmental Sciences,
DOI 10.1007/978-4-431-54838-6_14, © Springer Japan 2014

vermicomposting up to 60 days, clearly indicate the potential of earthworm biotechnology in recycling of waste, nutrient enrichment in the form of vermicompost.

Keywords Bioconversion • Biotechnology • Metaphire posthuma • Recycling

14.1 Introduction

With progress in industrialization and consequent urbanization, not only the quantity of solid waste increased, but its quality has also changed. The misuse and abuse of environment increased accumulation of wastes, impairing health and well being of human beings. In India, about 3,000 million tons of waste is produced annually. Piles of garbage and wastes of all kinds littered everywhere have become common sight in our urban life. All these waste materials are assimilated pollution and degradation. Tackling this huge quantity of wastes with currently used methods of treatment and disposal has met with little success and seems to be unsustainable. Around our dwelling, the effective and fruitful disposal of garbage is one of the prime problems which need to be taken up at priority. A number of solid waste disposal strategies have been adopted around the world. Most common practices of waste processing are uncontrolled open dumping, hog feeding, land filling, land spreading and incineration. Each solution has its own benefits and limitations.

An ecosystem cannot absorb them at natural cost and the pilling of such waste pose extra ordinary economic and environmental problems. The increasing waste generation rate, high collection cost and dwindling financial resources are the major problems faced by most of the developing countries for efficient solid waste management. Hence, there is an urgent need for a technique to save the future generation from adverse effects of solid waste. Solid waste management includes all activities that benefit public health in particular, environmental quality in general and minimizes the aesthetic impacts of solid waste directly and substantially. Unless organic solid waste is managed appropriately, its adverse impact continues contaminate the soil and environment. The fully decomposition and stabilization of these wastes through the biological treatment appears to be most cost effective and carry a less negative environmental impact.

Among the various possible ways to improve the nutrient status of organic solid waste, vermicomposting appears to be boon to the economy. Vermicomposting is ecotechnology rooted in principles of ecology, economics and equity. The basic aim of vermicomposting is to bring about decomposition of organic solid wastes without loss of nutrients and the production of end product rich in plant nutrients, which are used for agricultural and horticultural uses. The nutrients contained in the organic wastes are partly converted to more bioavailable forms in vermicomposting. In the present study an attempt has been made to manage the agricultural waste by vermicomposting using an indigenous earthworm *Metaphire posthuma*. The experiments were carried out for a period of 60 days to assess the ability of '*Metaphire posthuma*' to decompose the agricultural waste. Hence, this waste management biotechnology thus resulted in utilization of waste material into useful product on one side environment clean up another side.

14.2 Materials and Method

14.2.1 Collection of Material

Solid waste material was selected for the experiment, then air dried and grinded into smaller pieces. This grinded waste material was mixed with cow dung and was subjected aerobic composting to initiate microbial activity for 12 days. The pre composting is very essential to avoid the mortality of worms (Garg et al. 2006). The mixture was hand manipulated at regular time intervals and remoistened for sufficient activity. Specimens of Metaphire *posthuma* was obtained from laboratory stock.

14.2.2 Physicochemical Analysis

Waste samples were collected from the container by randomly mixing the waste and bedding soil. The sample dried at 110 °C for 5 h were passed through 2.0 mm sieve. During the composting process the material was analysed for different physiochemical parameter such as pH, organic matter, phosphorus as per standard methods (APHA 2005), total nitrogen by Kjeldahl method (Jackson 1973), potassium by flame photometric method (Simard 1993). During the experiment the samples were examined at regular interval of 15 days up to 60 days vermicomposting.

14.2.3 Experimental Design for Composting

The decomposition experiments were carried out in plastic container covered with muslin cloth to facilitate aeration in order to get final composted material. Worm worked compost was placed in each container to act as microbial inoculums and as a suitable habitat for the earthworms.

14.2.4 Statistical Analysis

Standard Deviation: $\delta = \sqrt{\frac{\Sigma y^2 - (\Sigma y^2)^2/N}{(N-1)}}$ where δ is the standard deviation; N is the number of observations.

Standard Error: The standard error of the mean is a measure of the reliability.

$$\text{SEM} = \frac{S}{\sqrt{N}} \text{ of the mean calculated from a set of observations.}$$

14.2.5 Three Way Analysis of Variance

A three-way analysis of variance (ANOVA) test of significance was used to evaluate the level of significance of difference between the vermicomposts produced by three substrates (cow dung, kitchen waste and agriculture waste) and control samples with respects to nutrient parameters. The data were analysed on software package statistica, version 7.0 programmes.

14.3 Results and Discussion

The physical and chemical parameters were changed in final vermicomposts with respect to initial feed substrate which revealed the potentiality of *Metaphire posthuma* in bioconversion of organic solid waste into nutrient rich vermicompost.

During vermicomposting the pH, electrical conductivity, organic matter, C: N ratio decrease till the end of the experimental tenure except total nitrogen, phosphorus and potassium. Lowest value of total nitrogen (0.93 ± 0.220), phosphorus (652.30 ± 128.761) and potassium (20.47 ± 4.037) were found at initial level. During vermicomposting period these parameters increases and their maximum values total nitrogen (0.98 ± 2.405), phosphorus (989.80 ± 117.690), and potassium (36.78 ± 3.615) were obtained after 60 days vermicomposting period.

pH of the raw waste before composting lies in alkaline range tend towards neutrality which might have been due to reduction of organic matter to mineral acids and thereby decreasing the alkalinity. The decomposition of organic matter produces 'organic acids' that lower the pH of the vermicomposts. With the advancement of composting process the decrease in pH of the compost was observed which can be attributed to the production of carbon dioxide, simple organic acid and loss of nitrogen as volatile ammonia at high pH values (Hartenstein and Hartenstein 1981) or mineralization of nitrogen and phosphorous into nitrates/nitrites and orthophosphate respectively by microbial activity (Ndegwa et al. 2000).

The decrease in E.C. at the end of compost formation was observed by Venkatrajan et al. (2001). Decrease in E.C. due to increased rate of loss of organic matter consequently release different minerals salt of this combination (Kaviraj and Sharma 2003). The decrease in organic matter may be due to that the carbon substrates are utilized by the microorganisms for respiration and for their cell growth. Earthworms and microorganisms uses large portion of carbon as source of energy (Venkatesh and Eevera 2008).

Final nitrogen content of vermicompost is dependent on the initial nitrogen content present in the organic wastes and the extent of decomposition (Crawford 1983). Earthworms enhanced the nitrogen levels of the substrate during digestion in their gut adding their nitrogenous excretory products, mucus, body fluid and enzymes in vermicomposting process (Suthar 2007). Nitrogen content is increased even through the decaying dead tissues of worms as proteineous portion was converted into ammonia and nitrogeneous like substances (Tripathi and Bhardwaj 2004).

During vermicomposting, the population of nitrogen fixer bacteria increased (Kavian and Ghatnekar 1991) which fix the nitrogen in the substrates. Earthworms strongly influence soil nitrogen status and nitrogen cycling, transferring nitrogen from decaying plant material to form that can be easily recycled and taken up by plants (Syers et al. 1979).

The unavailable forms of phosphorous transformed into the easily available form for plants through the action of worms during vermicomposting (Ghosh et al. 1999). Release of available phosphorous content from feed material may be due to the earthworm gut enzyme phosphatases and P-solubilizing microorganism present in worm casts (Suthar and Singh 2008). In the present study, it has been observed that potassium concentration was decreased in final vermicompost. The loss of potassium may be due to its higher susceptibility to leaching (Swift et al. 1979). Edwards and Burrows (1988) found that exchangeable potassium decreased in the worm worked substrate which may be due to greater leaching of potassium in comparison with other cations (Tables 14.1 and 14.2, Fig. 14.1).

After vermicomposting a decreases in potassium content due to leaching was reported by Garg et al. (2006). A decrease in potassium can be attributed to the microbial activities in the composting material as potassium is very much essential for their metabolic activity. The C: N ratio is the critical factor that limits earthworm's population. When the C: N ratio of the feed substrates increases, it becomes difficult to extract enough nitrogen for tissue production. C: N ratio play an important role in the nutrient balance in vermicomposting, this ratio told the amount of carbon available with respect to nitrogen for the composting microorganisms.

The reduction in C: N ratio was due to the fast degradation of organic matter mainly the degradation of cellulose and other readily available carbon and consequent volatilization of organic matter as the compost heats up. Microorganisms use carbon for both energy and growth while nitrogen is essential for protein production and reproduction. The nitrogen content in the composting material remained more or less same only it was transformed to inorganic forms, which might be the reason behind the drop in C: N ratio. Earthworms help to lower the C: N ratio of fresh

Table 14.1 Changes during vermicomposting of organic waste using *Metaphire posthuma*

Treatments				
Observations	Sets	pH	E.C.	Organic matter (per 100 g)
No. of days	Initial	8.14	0.64 ± 0.026	76.72 ± 2.805
15 days	Cont.	8.09 − (0.61)	0.61 ± 0.086 − (4.68)	71.91 ± 2.801 − (6.26)
	Expt.	7.98 − (1.96)	0.45 ± 0.158 − (29.68)	64.79 ± 2.971 − (15.55)
30 days	Cont.	8.10 − (0.49)	0.62 ± 0.266 − (3.13)	65.05 ± 4.667 − (15.21)
	Expt.	7.69 − (5.52)	0.38 ± 0.177 − (40.63)	55.78 ± 3.534 − (27.29)
45 days	Cont.	7.97 − (2.09)	0.59 ± 0.152 − (7.81)	64.03 ± 2.616 − (16.54)
	Expt.	7.47 − (8.23)	0.30 ± 0.707 − (53.13)	44.93 ± 1.413 − (41.43)
60 days	Cont.	7.91 − (2.83)	0.43 ± 0.125 − (32.81)	57.287 ± 0.960 − (25.32)
	Expt.	7.30 − (10.31)	0.22 ± 0.076 − (65.63)	36.78 ± 3.615 − (52.05)

Values are mean ± S.D., n = 6, Values in parentheses are % alteration

Table 14.2 Changes during vermicomposting of organic waste using *Metaphire posthuma*

Treatments				
Observations	Sets	Nitrogen (%)	Phosphorous (per 100 g)	Potassium (per 100 g)
No. of days	Initial	0.93 ± 0.220	652.30 ± 128.761	20.47 ± 4.037
15 days	Cont.	0.91 ± 0.498 − (2.15)	642.02 ± 31.714 − (1.60)	19.28 ± 2.430 − (5.81)
	Expt.	0.89 ± 2.261 − (4.30)	817.14 ± 85.499 + (25.27)	17.68 ± 2.180 − (13.62)
30 days	Cont.	0.89 ± 0.743 − (4.30)	622.44 ± 73.992 − (4.58)	18.37 ± 2.586 − (10.25)
	Expt.	0.87 ± 1.372 − (6.45)	876.90 ± 71.992 + (34.43)	14.17 ± 2.270 − (30.77)
45 days	Cont.	0.86 ± 1.432 − (7.52)	659.22 ± 116.771 + (1.06)	17.56 ± 1.618 − (14.21)
	Expt.	0.91 ± 3.540 − (2.15)	882.84 ± 67.510 + (35.34)	10.08 ± 2.958 − (50.75)
60 days	Cont.	0.83 ± 1.577 − (10.75)	662.48 ± 34.682 + (1.56)	16.46 ± 2.712 − (19.58)
	Expt.	0.98 ± 2.405 + (5.37)	989.80 ± 117.690 + (51.73)	36.78 ± 3.615 − (52.05)

Values are mean ± S.D., n = 6, Values in parentheses are % alteration

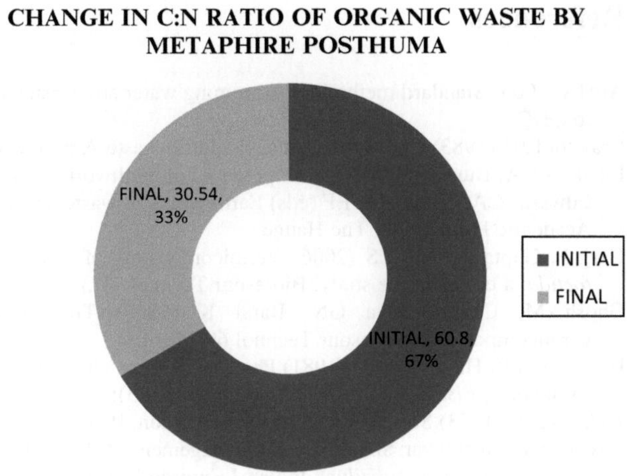

Fig. 14.1 Change in C: N ratio of organic waste

organic matter by consuming the matter, breaking it down and using the carbon for energy during respiration (Ronald and Donald 1997).

The ratio of carbon to nitrogen is important for the proper growth of any plant. All studies on vermicomposting have reported a decrease in C: N ratio of organic wastes although decrease in C: N ratio is different for different organic wastes. The increase in earthworm population might also be attributed to the C: N ratio decreasing with time (Ndegwa et al. 2000). C: N ratio was significantly lower in the treatments involving vermicomposting which indicates that they underwent more intense decomposition.

14.4 Conclusion

From the present study, It can be concluded that vermicomposting mediated by earthworms is an ecofriendly waste management technology and resulting in the bioconversion from waste to wealth. Composting by earthworm is proving to be environmentally preferred process over the normal microbial composting and much more over the landfills, as it is rapid and nearly odourless process, can reduce composting time significantly and there is no emission of green house gas methane. All-in-all, the vermicompost is believed to be very good organic fertilizer and soil conditioner.

References

APHA (2005) Standard method for examining water and wastewater, 21st edn. APHA, Washington, DC

Crawford JH (1983) Composting of agricultural waste-A review process. Biochem 18:14–18

Edwards CA, Burrows I (1988) The potential of earthworms composts as plant-grown media. In: Edwards CA, Neuhauser EF (eds) Earthworms in waste and environmental management. SPB Academic Publication, The Hauge

Garg P, Gupta A, Satya S (2006) Vermicomposting of different types of waste using *Eisenia foetida*: a comparative study. Bioresour Technol 97:391–395

Ghosh M, Chattopadhya GN, Baral K (1999) Transformation of phosphorus during vermicomposting. Bioresour Technol 69:149–154

Hartenstein R, Hartenstein F (1981) Physicochemical changes effected in activated sludge by the earthworm, *Eisenia foetida*. J Environ Qual 10(3):377–382

Jackson ML (1973) Soil chemical analysis, 1st edn. Prentice Hall, New Delhi

Kavian MF, Ghatnekar SD (1991) Biomanagement of dairy effluents using culture of red earthworms (*Lumbricus rubellus*). Indian J Environ Prot 11:680–682

Kaviraj P, Sharma S (2003) Municipal solid wastes management through vermicomposting employing exotic and local species of earthworm. Bioresour Technol 90:169–173

Ndegwa PM, Thompson SA, Das KC (2000) Effects of stocking density and feeding rate on vermicomposting of biosolids. Bioresour Technol 71(1):5–12

Ronald EG, Donald ED (1997) Earthworms for ecology and profit. In: Scientific earthworm farming, Bookworm Publishing Company, Ontario

Simard RR (1993) Ammonium acetate extractable elements. In: Martin R, Carter S (eds) Soil sampling and methods of analysis. Lewis Publisher, Boca Raton, pp 39–43

Suthar S (2007) Influence of different food sources on growth and reproduction performance of composting epigeic: Eudrilus engeniae, Perionyx excavatos and Perionyx Sansibaricus. Appl Ecol Environ Res 5(2):79–92

Suthar S, Singh S (2008) Vermicomposting of domestic waste by using two epigeic earthworms (Perionyx excavatus and Perionyx sansibaricus). Int J Environ Sci Tech 5(1):99–106

Swift MJ, Heal OW, Anderson JM (1979) Decomposition in terrestrial ecosystems. University of California Press, Berkeley

Syers JK, Sharpley AN, Keeney DR (1979) Cycling of nitrogen by surface casting earthworms in pasture ecosystem. Soil Biol Biochem 11:181–185

Tripathi G, Bhardwaj P (2004) Comparative study on biomass production, life cycle and efficiency of *Eisenia foetida* (Savigny) and Lampito mauritii (Kingberg). Bioresour Technol 92:275–283

Venkatesh RM, Eevera T (2008) Mass reduction and recovery of nutrients through vermicomposting of fly ash. Appl Ecol Environ Res 6(1):77–84

Venkatrajan B, Balalurugan V, Gobi M, Vijay Lakshmi GS (2001) Conversion of leaf litter through three different composting methods. J Ecophysiol Occup Hlth 1:59–66

Chapter 15
Impact of Organic Farming in Enhancing the Soil Microbial Pool

Sudha Sambyal Malik and Ramesh C. Chauhan

Abstract One of the important indicators of soil quality is the soil organic matter which influences the soil microbial population dynamics and enzyme activities, which in turn affect the soil fertility. Present study was conducted in the soil microbiological section, Department of Soil Science, CSKHPKV, Palampur in rice-lentil cropping sequence with organic, integrated and inorganic nutrient management. There were eight treatments with three replications and randomized block design. The surface and subsurface soil samples were collected before sowing and after harvesting from each treatment (0–15 and 15–30 cm) deep samples. The soil samples were analyzed to study how the microbial properties changed with changing the inputs in soil. An addition of organics along with in-organics improved urease and phosphatase activity in the soil. Whereas the application of only organic inputs was found to improve the soil biological properties such as microbial population, biomass carbon, microbial respiration and dehydrogenase activity. These microbial properties play a significant role in nutrient cycling, improvement of soil structure and many other functions, which directly and indirectly improve the soil health.

Keywords Integrated treatment • Microbial properties • Nutrient management • Organic treatment

---

S. Sambyal Malik (✉) • R.C. Chauhan
Department of Biology and Environmental Sciences, College of Basic Sciences,
CSK Himachal Pradesh Krishi Vishvavidyalaya, Palampur 176062, India
e-mail: sambyal.sudha@gmail.com

M. Singh et al. (eds.), *Climate Change and Biodiversity: Proceedings of IGU Rohtak Conference, Vol. 1*, Advances in Geographical and Environmental Sciences,
DOI 10.1007/978-4-431-54838-6_15, © Springer Japan 2014

15.1 Introduction

Health and environmental issues associated with the intensive use of chemical inputs has led to the interest in alternate forms of agriculture in the world. Due to the challenges from population growth, urbanization, and environmental degradation there is an increase in food insecurity in the world. Organic agriculture is one among the broad spectrum of production methods that are supportive for environment. This is done by using where possible agronomic, biological and mechanical methods, and opposing the use of synthetic materials to fulfill any specific function within the system. A form of agriculture which is mainly based on the use of organic fertilizers, natural pesticides, natural feed for cattle and poultry, and indigenous varieties of crops is an organic farming. Organic farming works in harmony with nature rather than against it. Organic agriculture is now practiced in more than 130 countries with a total area of 30.4 million ha in 0.7 million number of organic farm. This constitutes about 0.65 % of the total agriculture land of the world (Willer et al. 2008). In India, about 528,171 ha area is under organic farming with 44,926 numbers of certified organic farms (Ramesh et al. 2010).

Basically, in organic farming nutrient management is needed by creation of healthy and fertile soil. Nutrient management is the process of managing the amount, source, timing, and method of nutrient application with the goal of optimizing farm productivity while minimizing nutrient losses that could create environmental problems. The soil organic matter is key to good soil quality. Soil microbial property is one of the measures of biological activity and decomposition. It also provides an indication of soil's ability to sustain plant growth. A fertile soil has greater plant growth, which can create greater inputs of roots and other plant debris into the soil. This plant debris undergoes decomposition and adds to the soil organic matter. Applications of animal manures and composts, as well as the use of cover crops, all help increase soil organic matter. Organic matter provides a food source for soil microbes and increases microbial activity. As the microbes breakdown organic matter, nutrients are released in forms that the plants can utilize. Because nutrient management accounts for the nutrients added to the system, it promotes increasing soil quality without creating nutrient excesses. So the sustainable management of these natural resources for achieving food, nutritional, environmental and livelihood security in the country is most important.

The required capacity of the soil should be maintained by precise monitoring of changes in soil fertility and crop productivity in relation to proper nutrient management for sustainable production. As we know soil environment is greatly influenced by the different agricultural practices. Thus there is a need for the proper management of the agroecosystem for a sustainable agriculture. It was found that biological parameters of soil quality were generally enhanced in organic farming systems as compared to integrated systems (Fließbach et al. 2007). Thus, organic farming is the only farming system which can sustain and maintain health of agroecosystem and long term ecological and biological integrity of natural resources.

15.2 Material and Methods

15.2.1 General Description of the Area

15.2.1.1 Location

The research trial was conducted in soil microbiology section of CSKHPKV Palampur in rice-lentil cropping sequence with organic, inorganic and integrated nutrient management. The experimental farm is situated at 31°6/N latitude and 76°3/E longitude at an altitude of about 1,290 m above mean sea level. The site lies in the Palam valley of Kangra district in the mid hill sub humid zone of Himachal Pradesh.

15.2.1.2 Climate and Weather

The climate of the experimental site is characterized as wet temperate with mild summers (March to June) and cool winters. The mean annual rainfall around Palampur during 2010–2011 was 1,500–3,000 mm. The mean maximum temperature remains about 31 °C during the hottest months of May to June. December to February are the coldest months with mean minimum temperature of about 13.6 °C.

15.2.1.3 Soil

Soil of the study area at the start of the experiment was silty, clay loam in texture and classified as Typic Hapludalfs as per the Taxonomic system of soil classification (Soil Survey Staff 1975).

15.2.2 Field Studies

15.2.2.1 Experimental Details

The field experiment was conducted on a pre-established experiment which comprised of eight treatments. The treatments were:

T_1—10 tones Vermicompost + Bio-fertilizer + Chopped crop residues

T_2—NPK – 90:40:40 (50 kg Nitrogen will be substituted by 3.5 tones Vermicompost +67 kg/ha NeemCake + 20 kg/haBio-fertilizer)

T_3—5 tones Vermicompost + Bio-fertilizer + Half Nitrogen + Recommended P and K

T_4—5 tones Vermicompost + Bio-fertilizer + Half N and P + Recommended K
T_5—1.75 tones Vermicompost + Bio-fertilizer + 67 kg/ha Neem cake + Half Nitrogen + Recommended P and K
T_6—1.4 tones Vermicompost + Bio-fertilizer + Half N and P + Recommended K
T_7—Recommended doses of NPK-90:40:40
T_8—Control

Numbers of treatment were 8 (eight) and replications were 3 (three). Plot size was 3.5 × 3.0 m and research design was Randomized Block Design (RBD). Recommended dose of chemical fertilizer NPK for Rice-90:40:40 and Lentil-10:20:10. Fertilizer's Source was Urea, Single Super Phosphate (SSP) and Murate of Potash (MOP) and bio-fertilizers used were *Azospirillum* and *Phosphate solublizing bacteria*(PSB).

15.2.2.2 Soil Sampling

Surface and subsurface soil samples (0–15 cm) and (15–30 cm) were collected before sowing and after the harvest of the three cropping sequences and were air dried and grinded in a wooden pestle and mortar to pass through 2 mm sieve and subsequently stored in polyethylene bags for determination of biological parameters.

15.2.3 Laboratory Studies

The processed soil samples were analysed for microbial properties (total microbial count. biomass carbon, microbial respiration, enzyme activity i.e. phosphatase, dehydrogenase and urease) by following the standard methods. The enumeration of microbial population was done by plate count technique of (Wollum 1982) through serial dilution using a variety of media. Microbial biomass carbon was determined by fumigation-extraction method of Vance et al. (1987). The microbial respiration was determined by the method as described by Stotzky (1965). The dehydrogenase activity was determined by the method as described by Casida et al. (1964). For phosphatase activity method as described by Tabatabai and Bremner (1969) was used. The urease activity was determined by the method as described by Tabatabai and Bremner (1972). The data generated from the study was subjected to the statistical analysis through the requisite statistical computation following the procedure as outlined by Gomez and Gomez (1984).

15.3 Results and Discussion

15.3.1 Total Microbial Count

15.3.1.1 Bacterial Population

The bacterial population in soil at both the depths 0–15 and 15–30 was found to be increased in after harvesting soil samples. Amongst all the treatments the maximum value was recorded in treatment T_1 (10 tones vermicompost + biofertilizers + chopped crop residues) i.e. 6.89 cfu/g soil in after harvesting soil samples and 6.88 cfu/g of soil in before sowing samples˙ And the minimum 1.86 cfu/g of soil in treatment T_8 (control) in after harvesting soil samples and 1.83 cfu/g of soil in before sowing samples. The T1 was followed by integrated treatments (T_3 to T_6). Amongst all the three integrated treatments maximum bacterial population was found to be in T_6 (1.4 tones Vermicompost + Bio-fertilizer + Half N and P + Recommended K.) i.e. 6.34 cfu/g of soil in after harvesting and 4.66 cfu/g in before sowing soil samples (Table 15.1, Fig. 15.1). The bacterial population was found to be higher in 0–15 cm depth as compared to 15–30 cm depth mainly due to the high availability of organic matter on surface soil. The probable reason for higher population in T_1 is the high inputs of organic fertilizers added, maintaining the continuous supply of as well as energy for the growth of microorganisms.

The bacterial population were found to be consistently higher in organically managed soil than conventionally cultivated legume field (Chhotaray et al. 2010a, b). Microbial biomass and their count increase with increasing application of organic manure (FYM and others) and both the parameters are more in treatment receiving inorganic N along with FYM over a period of time due to better root growth, (Kukreja et al. 1991; Goyal et al. 1992; Sriramachanderasekhran et al. 1995).

Table 15.1 Effect of different treatments on bacterial population

Bacterial population ($\times 10^6$ cfu/g Soil)	Before sowing Depth (cm)		After harvesting Depth (cm)	
Treatments	0–15	15–30	0–15	15–30
T1	6.88	6.31	6.89	6.1
T2	6.13	5.64	6.59[a]	5.94[a]
T3	4.5	3.93	5.92	4.43
T4	4.03	3.31	5.19	4.58
T5	3.8	3.4	5.76	4.93
T6	4.66	4.5	6.34[a]	5.4
T7	2.83	2.13	2.94	2.41
T8	1.83	1.64	2.04	1.86
CD value($P = 0.05$)	0.69	0.52	0.76	0.55

[a]The mean is non significant

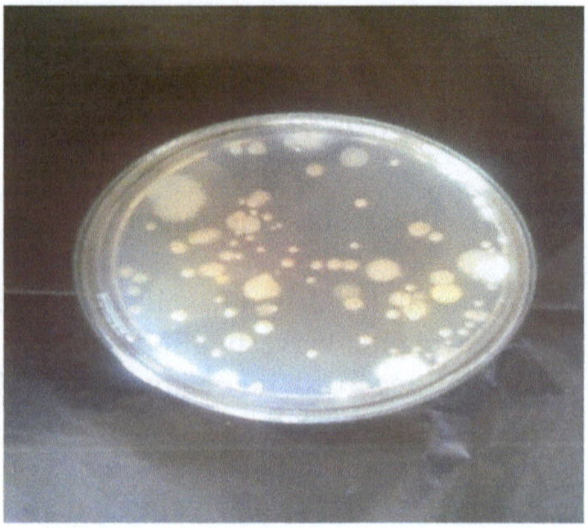

Fig. 15.1 Bacterial population in soil

Table 15.2 Effect of different treatments on fungal population

Fungal population ($\times 10^4$ cfu/g soil)	Before sowing Depth (cm)		After harvesting Depth (cm)	
Treatments	0–15	15–30	0–15	15–30
T1	4.88	4.23	4.95	4.75
T2	4.67[a]	4.33[a]	4.93[a]	4.29[a]
T3	4.15[a]	3.73[a]	3.9[a]	3.71[a]
T4	3.9[a]	3.82[a]	4.12[a]	3.46
T5	4.47[a]	4.1[a]	4.66[a]	4.69[a]
T6	4.29[a]	3.9[a]	4.63[a]	4.43[a]
T7	3.65[a]	3.67[a]	3.92	3.78[a]
T8	2.87	2.33	3.05	2.89
CD VALUE(P = 0.05)	1.29	1.22	1.27	1.21

[a]The mean is non significant

15.3.1.2 Fungal Population

The values were set at 10^4 dilution. The fungal population was found to lie between 2.87 and 4.95 cfu/g of the soil in surface samples and 2.33–4.75 cfu/g of the soil in subsurface samples. It was found to be significantly higher in T_1 (10 tonnes vermicompost + biofertilizers + chopped crop residues) in after harvesting samples i.e. 4.95 cfu/g of soil as compared to the minimum observed in control T_8 treatment i.e. 2.89 cfu/g of soil. The before sowing samples showed less population as compared to in after harvesting population. The fungal population was higher in surface soil (0–15 cm) as compared to subsurface soil (15–30 cm) in both the samples before sowing as well as after harvesting soil samples. The reason for higher population in T_1 could be the high inputs of organic fertilizers. The application of vermicompost could have increase the fungal population in T1 treatment (Table 15.2, Fig. 15.2).

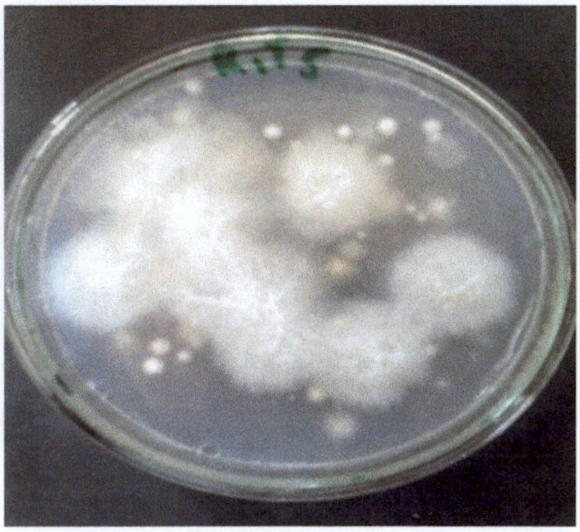

Fig. 15.2 Fungal population in soil

Table 15.3 Effect of different treatments on actinomycetes population

Actinomycetes population (× 10^3 cells/g Soil)	Before sowing Depth (cm)		After harvesting Depth (cm)	
Treatments	0–15	15–30	0–15	15–30
T1	60	48.33	63.66	50
T2	51	44[a]	57	48.88[a]
T3	47	43.66[a]	50	46.66[a]
T4	48	40[a]	55	47[a]
T5	41	37.66	48	40.33
T6	36	24	41.33	34
T7	35	23.66	45	34.66
T8	23	18	29.66	21.66
CD VALUE(P = 0.05)	6.65	8.64	5.02	5.99

[a]The mean is non significant

One study also indicated that counts of bacteria and fungi were higher in soils from organic in comparison to conventional farming systems (Chhotaray et al. 2010a).

15.3.1.3 Actinomycetes Population

The actinomycetes population ranged from 23 to 63.66 × 10^3cfu/g of soil in surface samples and it ranged from18 to 50 × 10^3 cells/g of soil in subsurface samples. The population of actinomycetes was found to be highest in the treatments where vermicompost was added. The vermicompost was added in organic treatments (T_1 and T_2) and in integrated treatments (T_3–T_6). Amongst all the treatments that had vermicompost T_1 of after harvesting gave the maximum population of actinomycetes i.e. 63.55 × 10^3 cells/g of soil. Whereas minimum value for actinomycetes population was observed in T_8 i.e. 29.66 × 10^3 cells/g of soil in 0–15 cm depth for after harvesting population (Table 15.3, Fig. 15.3). The population increased in after

Fig. 15.3 Actinomycetes population in soil

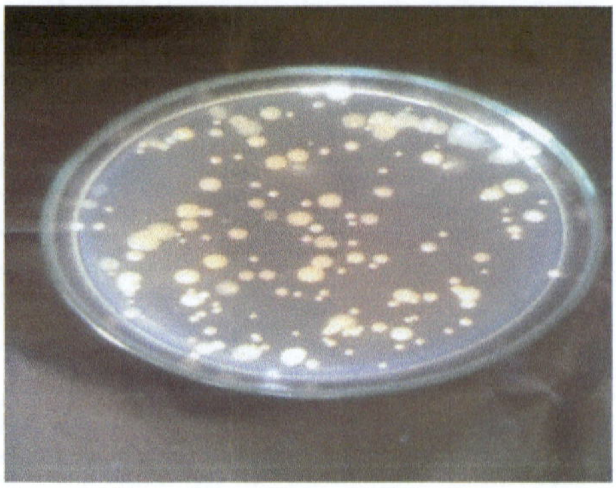

harvesting samples in the plots fed with Vermicompost, bio-fertilizers and organic manure treatments. Actinomycetes population was also higher in soils amended with organic amendments than soils amended with synthetic fertilizers (Bulluck et al. 2002).

In general the total as well as individual population was higher at surface soil in comparison to subsurface. Results are corroborated with the findings of Bedi and Dubey (2009), Bedi et al. (2009). Soil under organic agricultural system present higher microbial activity and biomass (Araujo et al. 2009). Organically treated plots have the maximum microbial population counts (Chauhan et al. 2011). Soil quality on conventional farms was significantly improved over a 2-year period by the addition of organic fertility amendments (Bulluck et al. 2002).

15.3.2 Microbial Respiration

The amount and rate of CO_2 released is an amount of metabolic activity of microorganisms present in the soil. The value of microbial respiration ranged from 5.56 to 13.46 μg/g/h in surface samples and 5.44 to 12.5 μg/g/h in sub surface samples (Table 15.4). The CO_2 evolved in control treatment was found lowest at both the depths in before sowing and after harvesting soil samples. It may be attributed to the low number of total microbial population in T_8. The maximum value was recorded in T_1 (13.46 μg/g/h) at depth 0–15 cm in after harvesting soil samples. Among the integrated treatmentsT4 (5 tones Vermicompost + Bio-fertilizer + Half N and P + Recommended K.) was found to have the maximum value of microbial respiration as compared to rest of the integrated treatments and chemical treatments. It might be attributed to the fact that microbial population was highest in this treatment. The same factors were found contributing for this by Liang et al. (2003). Surface soil showed more microbial respiration than subsurface soil because of high microbial population in the former than in the latter.

Table 15.4 Effect of different treatments on microbial respiration

Microbial respiration (μg/g/h)	Before sowing Depth (cm)		After harvesting Depth (cm)	
Treatments	0–15	15–30	0–15	15–30
T1	11.58	10.52	13.46	12.5
T2	11.46	10.50	12.47	11.55
T3	8.42	8.32	9.46	8.61
T4	11.50	10.70	13.18	12.60
T5	10.04	9.56	11.58	10.56
T6	9.52	8.76	10.58	10.16
T7	6.96	5.78	6.72	5.91
T8	5.56	5.44	6.17	5.62
CD value(P = 0.05)	1.26	1.58	1.44	1.48

Table 15.5 Effect of different treatments on dehydrogenase activity

Dehydrogenase activity (μg TPF/g/h)	Before sowing Depth (cm)		After harvesting Depth (cm)	
Treatments	0–15	15–30	0–15	15–30
T1	3.33	3.21	3.41	2.3
T2	2.45[a]	2.05	1.84	1.81[a]
T3	2.24	1.95	1.65	1.55
T4	1.94	1.82	1.71	1.06
T5	1.56	1.53	1.85	1.72[a]
T6	1.75	1.6	1.73	1.64[a]
T7	1.27	1.52	1.56	1.32
T8	1.19	1.25	1.46	1.38
CD value(P = 0.05)	1.01	0.77	0.85	0.71

[a]The mean is non significant

15.3.3 Enzymatic Activities

Soil enzymes play an important role in organic matter decomposition and nutrient cycling and its mobilization to different soil strata (Pavel et al. 2004; Shi et al. 2006).

15.3.3.1 Dehydrogenase Activity

Biological oxidation of organic matter is generally through the process of dehydrogenation. The enzymes dehydrogenase transfer hydrogen from substrate to acceptors. The dehydrogenase activity was lowest in control. It might be due to the fact that organic carbon level was also lowest in control. The maximum value was obtained in T_1 treatment. (10 tonnes vermicompost + biofertilizers + chopped crop residues) The dehydrogenase activity in all treatments was higher in surface soil than subsurface soil (Table 15.5). It might be due to the maximum organic

Table 15.6 Effect of different treatments on phosphatase activity

Phosphatase activity (μg/g/h)	Before sowing Depth (cm)		After harvesting Depth (cm)	
Treatments	0–15	15–30	0–15	15–30
T1	1.86	1.22	3.27	2.34
T2	1.96	1.57[a]	3.63	2.96
T3	1.55	1.65[a]	3.67[a]	3.34[a]
T4	1.76	1.65[a]	3.86[a]	3.73[a]
T5	2.35	1.86[a]	3.97[a]	3.12[a]
T6	2.96	2.21	4.36	3.95
T7	1.26	1.14	2.36	2.2
T8	1.16	1.08	2.19	1.53
CD value(P = 0.05)	0.39	0.76	1.04	0.80

[a]The mean is non significant

matter accumulation at soil surface. An Increase in its values in organic treatments could be due to the increased concentration of substrate. The higher dehydrogenase activity in surface soils as compared to subsurface soils may be ascribed to the fact that the maximum organic matter accumulation was at surface soil. Low in T_7 treatment may be due to the addition of chemical fertilizers which resulted in an accumulation of nitrate which inhibited the enzyme activity through interfering in the process of electron acceptor thus blocking the enzyme activity. The same results have also been reported by Goyal et al. (1992).

15.3.3.2 Phosphatase Activity

Its maximum value was obtained in treatment T_6 (1.4 tones Vermicompost + Bio-fertilizer + Half N and P + Recommended K) and the minimum in T_8. At sub-surface (15–30) soil this enzymatic activity was lower than the surface soil (0–15). The value increased from in 1.53 μg/g/h T_8 control to 4.36 μg/g/h in T6 (1.4 tones Vermicompost + Bio-fertilizer + Half N and P + Recommended K) (Table 15.6). The treatments which received the combined application of organics and inorganics (T_3 toT_6) gave higher phosphatase activity than others. It may attributed the addition of organic and inorganic sources which maintain the continuity of addition of nutrients, so the substrate of phosphorus i.e. monoesters or diesters are continuously available and cause phosphatase activity. Similar results were also obtained by Liang et al. (2003) in some soils from China. Lower values in complete organic treatments receiving chopped crop residues may be ascribed to narrow C: N ratio which resulted in slow mineralization from organic forms to di and mono esters. Combined use of organic manures improved the enzyme activity of the soil rather than single organic manure application (Krishankumar and Jawahar 2005).

Table 15.7 Effect of different treatments on urease activity

Urease activity (μg UREA/g dry soil/Min)	Before sowing Depth (cm)		After harvesting Depth (cm)	
Treatments	0–15	15–30	0–15	15–30
T1	3.56	2.95	3.3	2.74
T2	4.07	3.36	2.23	2
T3	3.42	3.12	3.48	3.07
T4	6.13	4.77[a]	5.49[a]	5.2
T5	5.24	4.60	4.58	3.8
T6	6.79	5.44	6.7	4.18
T7	3.52	2.5	2.37	2.15
T8	2.36	2.43	1.61	1.36
CD value(P = 0.05)	1.12	0.68	1.28	0.93

[a]The mean is non significant

15.3.3.3 Urease Activity

The values of urease activity ranged between 1.61 and 3.56 μg UREA/g dry soil/min in surface samples and 1.36–2.95 μg UREA/g dry soil/min in sub surface samples. Its maximum values were obtained in treatment T_6 (1.4 tones Vermicompost + Bio-fertilizer + Half N and P + Recommended K). The minimum value was obtained for T_8 (control). All integrated treatments were found superior to organic treatments followed by chemical and control (Table 15.7). The higher urease activity in T6 treatment may be attributed to the application of organic source and inorganic source together which maintain the continuity of conversion of nutrients from organic to inorganic form because it acts on C–N bond other than linear peptide bond.

15.3.4 Microbial Biomass Carbon

Fumigation extraction method gave lower values of microbial biomass carbon in acid soils, the food and energy(ATP) are derived from the same pool of organic matter source (Vance et al. 1987). The data in Table 15.8 revealed that the microbial biomass carbon was recorded less in control because of less microbial population. Microbial biomass accounts for the total carbon content contributed by living organisms in soil. The values varied between 107.37 and 383.03 μg/g in surface soil (0–15 cm) depth and from 114.80 to 288.51 μg/g in subsurface soil (15–30 cm). The maximum value (382.03 μg/g) was obtained in T1 treatment as compared to minimum (107.37 μg/g in) T_8, the control treatment. The microbial biomass carbon in before sowing samples gave lower values as compared to in after harvesting samples. The higher microbial population and organic carbon content was observed higher in these treatments which could have increased the biomass

Table 15.8 Effect of different treatments on microbial biomass carbon

Microbial biomass carbon (μg/g)	Before sowing Depth (cm)		After harvesting Depth (cm)	
Treatments	0–15	15–30	0–15	15–30
T1	287.1	234.9	383.03	288.51
T2	268.7	232.03	341.73[a]	249.51
T3	207.43	187.3	259.45	212.58
T4	164.36	147.13	301.33	177
T5	237.93	306.8	291.3	244.07
T6	145.49	123.43	221.08	186.42
T7	138.67	115.83	126.8	155.73
T8	119.96	114.78	107.37	114.80
CD VALUE(P = 0.05)	17.95	13.66	33.55	46.14

[a]The mean is non significant

carbon in T_1 (10 tonnes vermicompost + bio-fertilizers + chopped crop residues) treatment. The similar findings have also been recorded by Goyal et al. (1992). Microbial biomass and activities were enhanced in organic systems emphasizing the important role of element cycling processes that are supported by an abundant and active soil biological community (Fließbach et al. 2007). Soil under organic agricultural system presents higher microbial biomass carbon (Chauhan et al. 2011).

15.4 Conclusions

The maximum microbial population was observed in the organic treatments followed by the integrated, in-organic and control treatments. The value of microbial respiration ranged from 5.44 to 13.46 μg/g/h (Table 15.4). The Microbial respiration was found to be increased with an increase in organic matter. The maximum dehydrogenase activity was recorded in organic and minimum in the control treatment. It ranged from 1.19 to 3.41 μg TPF/g/h (Table 15.5). The maximum value for the phosphatase activity was found in the integrated treatments (Table 15.6). The different treatments revealed a significant effect on urease activity of the soil. Its values varied between 1.36 and 6.79 μgUrea/g dry soil/min (Table 15.7). Maximum urease activity was shown by the integrated treatments followed by organic treatments. The significant increase in microbial biomass carbon has been recorded during the study with an increase in organic matter content. The microbial biomass carbon content decreased in the control treatment. It ranged from 107.37 to 383.03 μg/g (Table 15.8). The maximum values were obtained in the organic treatments followed by integrated, inorganic and control treatments.

It can be inferred from the present investigation that an addition of organics along with in-organics improved phosphatase and urease activity. Whereas the application of only organic inputs was found to improve the soil biological properties such as microbial population, organic matter, biomass carbon, microbial respiration etc. Thus, it can be said that the organic farming has a sizeable impact on the structural functioning of agro-ecosystem and this is the only farming system which can sustain and maintain the ecology and health of agro-ecosystem.

References

Araujo ASF, Leite LFC, Santos VB, Carneiro RFV (2009) Soil microbial activity in conventional and organic agricultural systems. Sustainability 1:268–276

Bedi P, Dubey YP (2009) Long-term influence of organic and inorganic fertilizers on nutrient build-up and their relationship with microbial properties under a rice-wheat cropping sequence in an acid Alfisol. Biomed Life Sci 57(3):297–306

Bedi P, Dubey YP, Naveen D (2009) Microbial properties under rice-wheat cropping sequence in an acid Alfisol. J Indian Soc Soil Sci 57(3):373–377

Bulluck LR, Brosius M, Evanylo GK, Ristaino JB (2002) Organic and synthetic fertility amendments influence soil microbial, physical and chemical properties on organic and conventional farms. Appl soil Ecol 19:147–160

Casida LE, Klein DA, Sautor T (1964) Soil dehydrogenase activity. Soil Sci 98(2):371–376

Chauhan PK, Singh V, Dhatwalia VK, Abhishek B (2011) Physico-chemical and microbial activity of soil under conventional and organic agricultural systems. J Chem Pharm Res 3(3):799–804

Chhotaray D, Mohapatra PK, Mishra CSK (2010a) Macronutrient availability and microbial population dynamics of soils under organic and conventional farming of legume crop. Int Quarterly J Life Sci 3:643–650

Chhotaray D, Mohapatra PK, Mishra CSK (2010b) Soil macronutrient availability and microbial population dynamics organic and conventional agroecosystems. Eur J Biol Sci 3(2):44–51

Fließbach A, Oberholzer HR, Gunst L, Mader P (2007) Soil organic matter and biological soil quality indicators after 21 years of organic and conventional farming. Agr Ecosyst Environ 118:273–284

Gomez KZ, Gomez AA (1984) Statistical procedure for Agricultural Research. John Wiley and Sons, New York

Goyal S, Mishra MM, Hooda IS, Singh R, Beri V, Chaudhary MR, Sidhu PS, Pashricha NS, Bajwa MS (1992) Build up of microbial biomass with continuous use of inorganic fertilizer and organic amendments. Proc Int Symp Nutrient Manage Sustained Prod 2(2):149–151

Krishankumar S, Jawahar D (2005) Coir pith compost. Kisan World 28(6):65

Kukreja K, Mishra MM, Dhankar SS, Kapoor KK, Gupta AP (1991) Effect of long manorial application on microbial biomass. J Indian Soc Soil Sci 39(4):685–688

Liang YC, Yang YF, Yang CG, Shen QR, Zhou JM, Yang LZ (2003) Soil enzymatic activity and growth of rice and barley as influenced by organic manure in an anthropogenic soil. Geoderma 115(3):149–160

Pavel R, Doyle J, Steinberger T (2004) Seasonal patteren of cellulose concentration in desert soil. Soil Biol Biochem 33:347–350

Ramesh P, Panwar NR, Singh AB, Ramana S, Yadav SK, Shrivastava R, Subba RA (2010) Status of organic ferming in India. Curr Sci 98(9):1190–1194

Shi W, Dell E, Bowman D, Iyyemperumal K (2006) Soil enzyme activities and organic matter composition in a turfgrass chronosequence. Plant Soil 288:285–296

Soil Survey Staff (1975) Soil taxanomy: a basic system of soil classification for making and interpreting soil surveys. US Department of Agriculture/ Soil Conservation Service US. Government Printing Office, Washington, DC

Sriramachanderasekhran MV, Mithusomy P, Ravichandran M (1995) Long term effects of fertilizers on microflora and soil properties in integrated inception. Ann Agr Res 16(2):195–200
Stotzky G (1965) Microbial respiration. In: Black CA (ed) Methods of soil analysis. Part II. American Society of Agronomy Inc, Madison, pp 1550–1569
Tabatabai MA, Bremner JM (1969) Use of p-nitrophenyl phosphatase for assay of soil phosphatase activity. Soil Biol Biochem 1(2):301–307
Tabatabai MA, Bremner JM (1972) Assay of urease activity in soils. Soil Biol Biochem 4:479–487
Vance ED, Brookes PC, Jenkinson DS (1987) An extraction method for measuring soil microbial biomass carbon. Soil Biol Biochem 19(2):703–707
Willer H, Yussefi-Menzler M, Sorensen N (2008) The world of organic agriculture statistics and emerging trends. Curr Statis 23–46
Wollum AG (1982) Cultural methods for soil microorganisms. In: Method of soil analysis. Part 2. Chemical and microbiological properties, 2nd edn. American Society of Agronomy, Inc. Soil Science Society of America, Inc. Publisher, Madison, pp 781–813

Chapter 16
Risk of Bio-diversity Degradation in North-East India and Prospect of Management with Geoinformatics: A Case Study of Assam

N. Devi, Anil Boro, and R. Barman

Abstract The North-Eastern part of India has been one of the richest regions of the world in respect of bio-diversity. The geographical location, physiographic and climatic characteristics of this region have together favoured the existence of rich, large and resourceful diversity of flora and fauna. The recent trend of human encroachments, followed by degradation through natural calamities of flood, soil erosion, etc. have been creating the risks of existence of environment and the species as well in the region. In the world map, this part of India's north-east covers only 0.05 % (equivalent to 255,036 km^2) of geographical area between 20°N and 29°30′N parallels and 89°46′E and 97°30′E meridians.

The integrated tools of Geoinformatics provided facilities for accurate mapping, management of large volume of data, capabilities of complex geo-spatial problems analysis, etc.

Assam is one of the states in North East India consisting naturally of micro variations in topography, climate, soil, flora and fauna, etc. Here, exist famous national parks, viz. Kaziranga, Manas, Nameri, Dibru-Saikhowa, and Orang. In addition to that the state has a large number of wildlife sanctuaries and along with some wildlife protected areas. These parks, sanctuaries and protected areas in the state have been confronting threats because of rapid population growth, encroachment and unplanned space management.

N. Devi (✉)
Formerly Faculty, Department of Geography, Pandu College, Guwahati, Assam, India
e-mail: nami.devi@rediffmail.com

A. Boro
Tihu College, Tihu, India
e-mail: anil.boro2@rediffmail.com

R. Barman
Department of Geography, Gauhati University, Guwahati, India
e-mail: dr.ratneswar_barman@rediffmail.com

M. Singh et al. (eds.), *Climate Change and Biodiversity: Proceedings of IGU Rohtak Conference, Vol. 1*, Advances in Geographical and Environmental Sciences,
DOI 10.1007/978-4-431-54838-6_16,

In this paper an attempt has been made to highlight the current scenario of the degradation of Biodiversity in North-East India and application of Geoinformatics for problem solving in the case study area of Assam.

Keywords Bio-diversity • Degradation • Encroachments • Fauna • Flora • Geoinformatics • Space management

16.1 Introduction

The North-Eastern part of India is one of the richest regions in the world in respect of Bio-diversity. This part of India's north east has a large number of National Parks, wildlife sanctuaries along with some protected areas. The recent trend of population growth, conversion of forestland for human use, felling of trees, poaching of animals, floods, etc. have been creating obstacles, and risk for protection and of existence for long of the rich bio-diversity of the region. These anti-environmental activities have been indicating as the process of degradation and extinction of a large number of flora and fauna including the one-horned Rhinos of Assam.

Different measures like, seminars, workshops and awareness camps, procession have been done and investment of funds allocated in the name of afforestation and wildlife protection and preservation in this part of the country. Even as these have been essential, well–organised monitoring and management system, with well-equipped modern tools and methods are yet to be introduced to protect and preserve these rich resources and bio-diversity of the region.

Geoinformatics including the concepts and techniques of remote sensing, GIS and GPS as the set of sophisticated tools and technique extends the capabilities of monitoring and analysing the degraded development of bio-diversity which may help for protection and management of these rich resources.

16.2 Objective of the Work

The main objective of the work is to highlight the scenario of biodiversity in this part of the globe and to discus the prospect of biodiversity management using Geoinformatics.

16.3 Database and Methodology

The secondary data for the purpose of the study are collected from the records and writings of different government and non Government organizations, Satellite Imagery (mainly from Google earth), LULC, (land use land cover) map prepared by NRSA based on Satellite Image, Toposheets of India published by Survey of India and US

Army Map Service (R.F. 1:250,000) as well as primary information gathered through field observations. DN Garmin has been used for data conversion and ArcGIS is used for mapping, analysing and calculations of different form of data.

16.4 Bio-diversity of North-East India

The region of Northeast India is largely covered by hilly terrain. Geographical location, physiographic characteristics as well as climatic capabilities in the region have favoured the growth and existence of rich flora and fauna. The region is one of the Biodiversity hotspot areas in the world (Fig. 16.1).

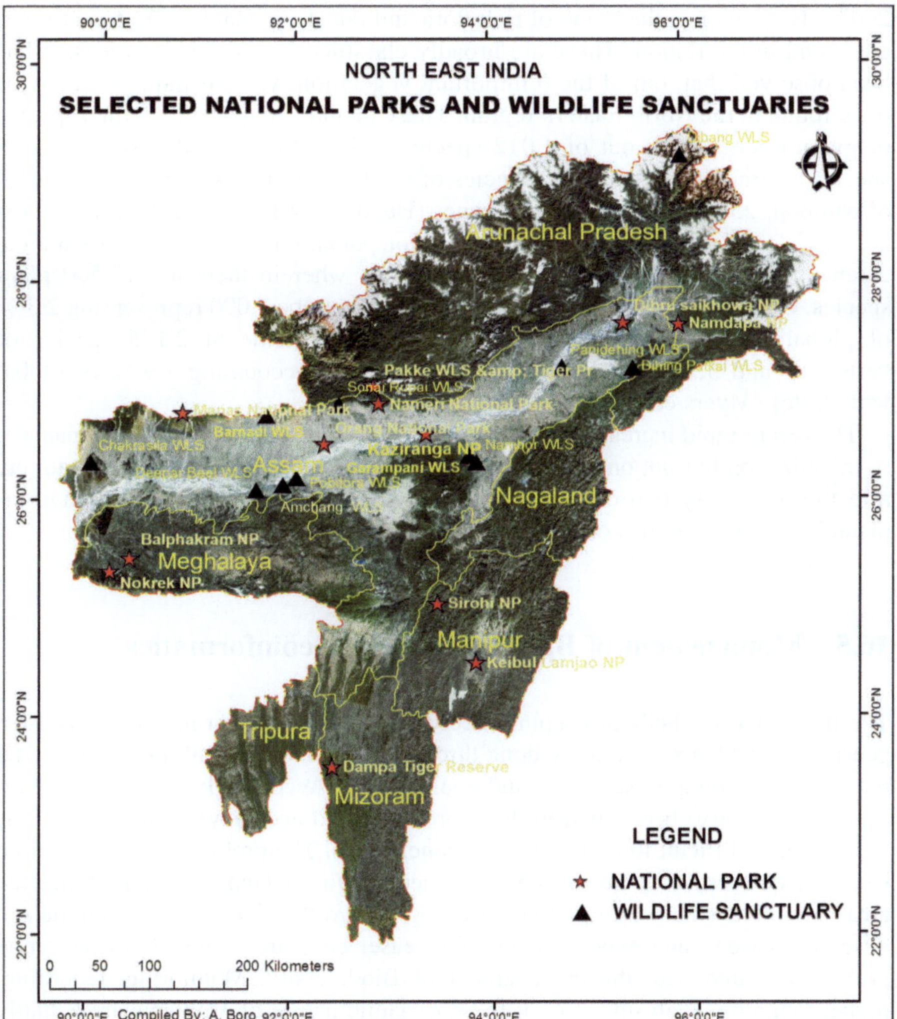

Fig. 16.1 National parks and wildlife sanctuary in N-E India

A large number of spots with extensive areas are declared as the National Parks and Wildlife sanctuaries by the governments of the states to protect concentrated flora and fauna in these areas. The National Parks in this part of the country are Kaziranga, Manas, Nameri, Dibru-Saikhowa, Orang/Rajib Gandhi in Assam; Namdapha, Mouling etc. National Parks in Arunachal Pradesh; Lamjao and Sirohi National Parks in Manipur, Balpakram and Nokrek National Parks in Meghalaya; Phawngpui and Murlen National Parks in Mizoram, and Intanki National Parks in Nagaland. The region, has also the wildlife sanctuaries, viz. the Fakim in Nagaland, Gumti and Sepahijala in Tripura, etc. In addition to these there are large numbers of wildlife sanctuaries and protected forest areas in this part of the country. North-East India has 173,297 km^2 of forest area accounting for 66.098 % of the total geographical area the region (State of Forest Report 2003). The region is the home of rich flora and fauna. As many as 51 forest types are found in the region. These are broadly classified into six major types. It has been observed that, out of the 9 important vegetation types of India, as many as 6 are found in the North Eastern region. These include 40 out of 54 Indian species of gymnosperms; 500 out of 1,012 species of Pteridophytes; 825 out of 1,145 species of orchids; 80 out of 90 species of rhododendrons; 60 out of 110 species of bamboo; 25 out of 56 species of canes (Hegde 2000). The North-East India is actually a contiguous part of Indo-Burma Hotspot and in the region original areal extent of primary vegetation is 2,060,000 km^2 wherein there are 13,500 plant species. Out of these endemic plants are estimated to be 7,000 representing 2.3 % of global plants. The region is further more the home of 2,185 species of vertebrate and 528 species of endemic vertebrates accounting 1.9 % of global vertebrates (Myers et al. 2000).

The recent rapid increase of human population followed by the rapid expansion of urbanization has not only been reducing the forest coverage but also negatively affecting the ecosystem of the region rendering intensifying impact on environmental natural calamities of flood, landslide, soil erosion, etc.

16.5 Management of Biodiversity and Geoinformatics

The traditional methods of keeping records as well as monitoring and analysis of geospatial problems are mainly done through manual works with the help of data recording, mapping, observation and analysis, etc. Accordingly methods of management have also been far from high precision and accuracy. As a result, it has become very difficult to more accurate conclusions. The problems of soil erosion; siltation; deforestation; human encroachment of forest land, wetlands, protected areas, etc.; river channel migration; poaching; road/rail incidence of wild animals in different seasons and time; spread of disease, etc., are some of the common problems related with the management of Biodiversity. Monitoring, recording, proper mapping, analysing, etc. of these dynamic nature of problems are actually a difficult task.

The difficulties of analysis of spatial characteristics, and are now overcome with the innovation and application of tools and techniques of Geoinformatics. The advantages of Geoinformatics cover the capabilities of availing and handling of large volumes of varying forms of data while accurately representing then and analyzing the complicated problem. Capabilities of data collection and generation in suitable temporal and spatial scales with the use of remote sensing, field work for ground information collection and checkup with the help of GPS and the storing, compiling, analyzing, etc., at GIS environment are the unique functions at the multi disciplinary applications of these tools and techniques. The nature of spatial data and the applicability and capability of the advantages, of Geoinformatics are represented in different works and writings. The examples like the basic knowledge of remote sensing and image interpretation (Lillesand and Keifer 2004) the fundamentals of GIS, the generation of spatial data and their analysis with GIS, project design with GIS, etc (Heywood et al. 2003); modeling with the GIS (Bonham-Carter 1994); Spatial Analysis Using GIS (Johnston and McCoy 2001); 3D visualization of landform (Booth 2000) etc. bear a great significant in solving difficult problems in the spatial contact.

16.6 Geoinformatics and Prospects of Biodiversity Management in Assam

Assam as a part of North East India has a legacy of significantly rich spots and areas of bio-diversity. There are a number of hot spots of this kind in the state. The state has 26.256 % of its total area under forest land (excluding un-classed State forest). The state registers five National Parks (NP) and more than dozens of Wildlife Sanctuaries. As per government records Assam has an area of 3,924.37 km^2 as National parks and wildlife sanctuaries (WLS) which accounts for 5.003 % of the geographical area of the state (Government of Assam 2009). According to the NRSA image based LULC map, forest (including all kinds forest, grassland, mixed forest with rural built up and shifting cultivation-abandoned area) cover accounts for 46.18 % while wetland, cover 2.27 % of the geographical area of the state. The National Parks in Assam including Kaziranga NP have been confronting problems for existence due to poachers and human encroachments for settlement and other kinds of landuse. In addition to that road and rail networks through the National parks and WLS along with the natural calamities of floods have caused growing incidences of animals.

As the case study of Assam, Kaziranga NP is selected specially for layers generation and explanation of problems. For the better solution of existing problems of Biodiversity management the following flow chart (Fig. 16.2) can be applied.

The common threat and information required for risk management of biodiversity (Table 16.1) are in front of Government and society. To overcome the risk and problems the management works have complicated day by day and in this situation

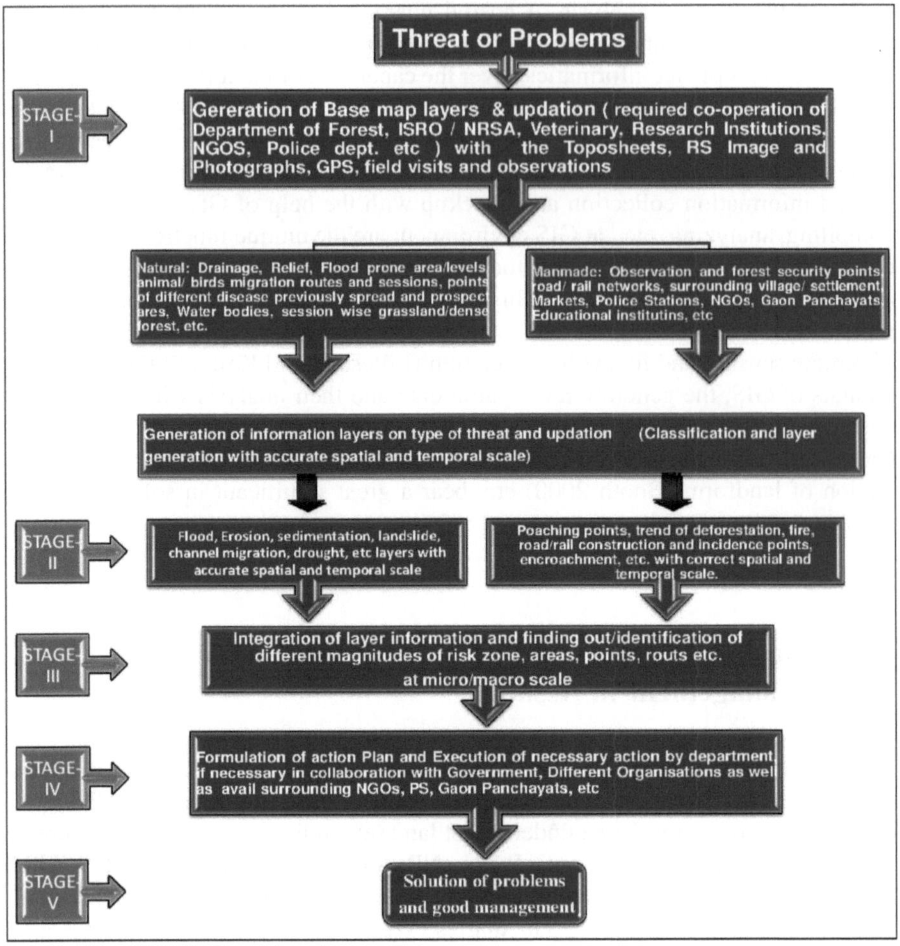

Fig. 16.2 Flow chart for management of biodiversity

in addition to prior experience, adaptation of new strategies including the tools and techniques of Geoinformatics and its necessary training are becoming as an essential alternative. The generation of required information map; preparation of maps and their updation along with the; monitoring; analysis and identification of risk points, and areas, etc. have become very effective. The generated layers of information that stand for the Kaziranga NP are presented for the solution of identified problem of Biodiversity management of the national park (Figs. 16.3 and 16.4)

To overcome or to have the solution of these problems compilation of threat-wise required and accurate information in different layers becomes prerequisite. The boundary delineation of risks and their analysis has also been difficult because natures of risks are changing over time. Sources of risk of biodiversity degradation boundaries may not within the outer zone of NP or WLS, rather it may relate to local, national, international or even seasonal dimensions.

Table 16.1 Common threat and the important information for risk management of Biodiversity specially in the Kaziranga NP

The common threat that arise to degradation of Biodiversity NE India and including Assam	The common information to be collected for the risk management of Biodiversity in many part of the world including NE India and Assam
In addition to climate change and the impact of global warming the common threats are- i. Human encroachment to protected forest and forest land in various forms. ii. Poaching of wild animals including endangered species. iii. Natural calamities of flood, cyclone, siltation, etc. iv. Incidence of wild animals in roads, railways and electric ware. v. Pollution from industries vi. Spread of disease through migratory birds. vii. Disturbance of food chain and ecosystem through natural and manmade calamities. viii. Avoidance by concerned authorities, lack of security measures and Biodiversity friendly laws, roles of corrupted officials, lack of awareness by surrounding inhabitants of forest about the need of conservation of biodiversity, etc.	In addition to data and maps on climate, common important Natural and Cultural information are- i. Contour, spot height, relief and soil maps. ii. Maps of flood and flood level at various points of time and stages. iii. Census records and updation of existing animals. iv. Records and maps of migratory birds in respect of census seasonal routes of movement. v. Animal security points and area and movement route maps during flood and normal times. vi. Maps indicating Land Use and Land Cover (LULC) at appropriate and precise scale (streams, water bodies, grassland, etc .are covered here. vii. Trend/Change of LULC in and around the protected areas at spatial and temporal scales. viii. Maps of all kinds of roads/railways connected with the protected areas. ix. Precise map and information of the surrounding villages, Gaon Panchayats, NGOs, Police Stations, Educational and other institutions, etc. x. Records and map of previous poaching points, areas, seasons, nature, etc. xi. Records and map of peoples involved in previous poaching, their movement, markets of poaching wild animals, etc. xii. Map indications of own (departmental) security points, their range and areas/points of security laps. xiii. Map indicating human encroachment in different forms (e.g. construction, tree cutting, grassing domesticated animals, etc.) and risk areas of encroachment. etc.

Source: Based on personal observation and consultation of some writing

16.7 The Selected Problem and the Kaziranga NP

The example of selected problem is to identify best observation point and areas (or the security points and areas) within the Kaziranga National Park for the construction of security points and security road connectivity.

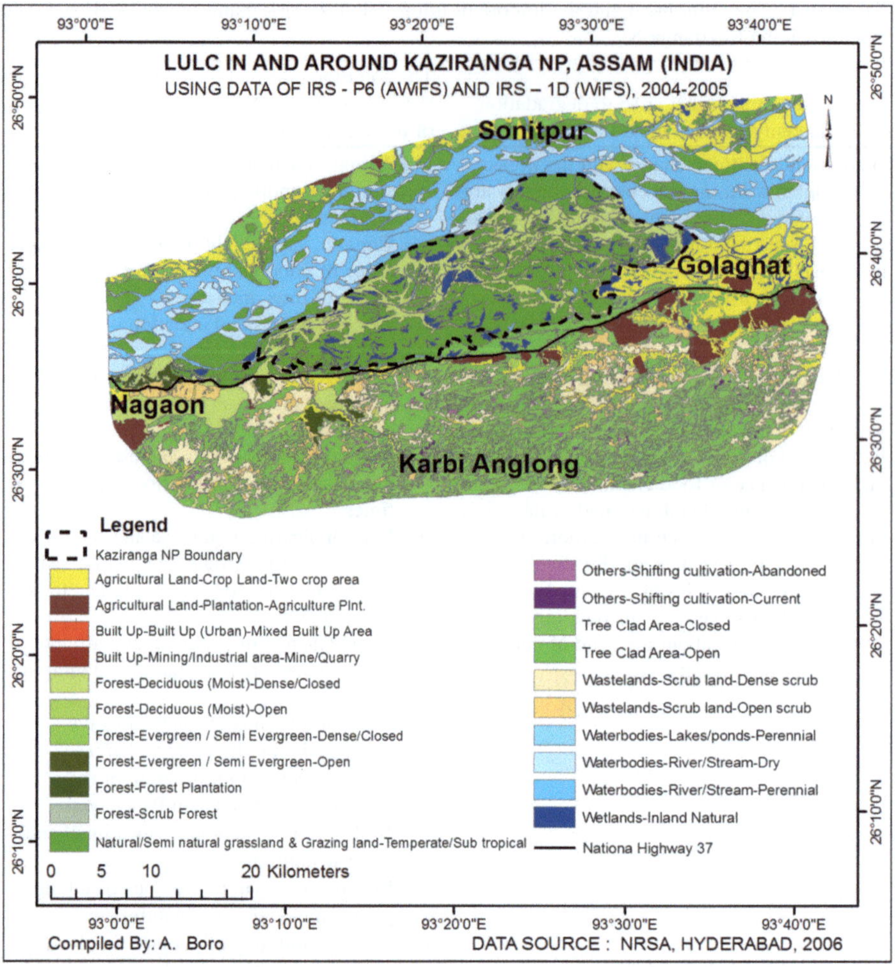

Fig. 16.3 LULC, in and around Kaziranga NP

The core area of the Kaziranga NP (covering an area of 430 km^2) is found in the areas between the Brahmaputra River and Karbi Anglong Plateau. This Kaziranga NP is bordered by the districts of Golaghat, Karbi Anglong, Nagaon and Sonitpur of Assam (Fig. 16.3). The centre of the NP lies at point of 93°25′E longitude and 26°40′N latitude. According to the UNESCO – IUCN Enhancing Our Heritage Project (2003) Current stress and threats of the NP are Poaching of wild animals, high flood, river bank and soil erosion, sedimentation and growth of weeds, illegal fishing heavy bus, motor car and truck traffic, live-stock grazing and the breach of embankments on eastern boundary of the Park. This results in unusual sudden increase of water level in the south of the park. Pollution and contaminations of the effluents oozing out from the oil industry and the intensified organised poaching have also their increasingly potential and effective threats to the bio-diversity of the Kaziranga NP.

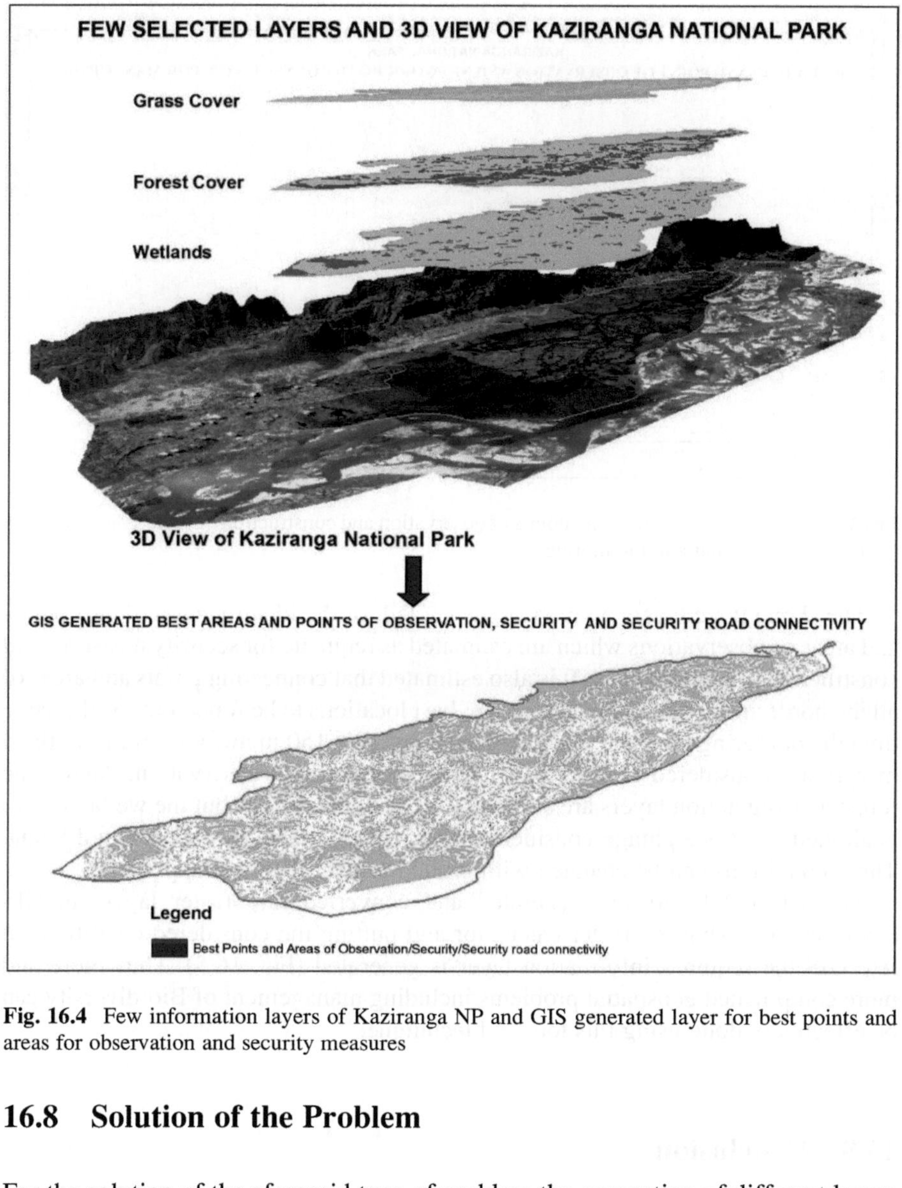

Fig. 16.4 Few information layers of Kaziranga NP and GIS generated layer for best points and areas for observation and security measures

16.8 Solution of the Problem

For the solution of the aforesaid type of problem the generation of different layers of LULC, boundary demarcation, contour/relief maps, identification of risk points of poaching, etc have been important. Here, three layers are generated and justified for the work, of course addition of more and more layers may give better result.

The layers are-

1. Dense forest cover map (Deciduous moist Dense/closed) of the Kaziranga NP
2. Grass cover map of the Kaziranga NP
3. Wetlands cover of Kaziranga NP

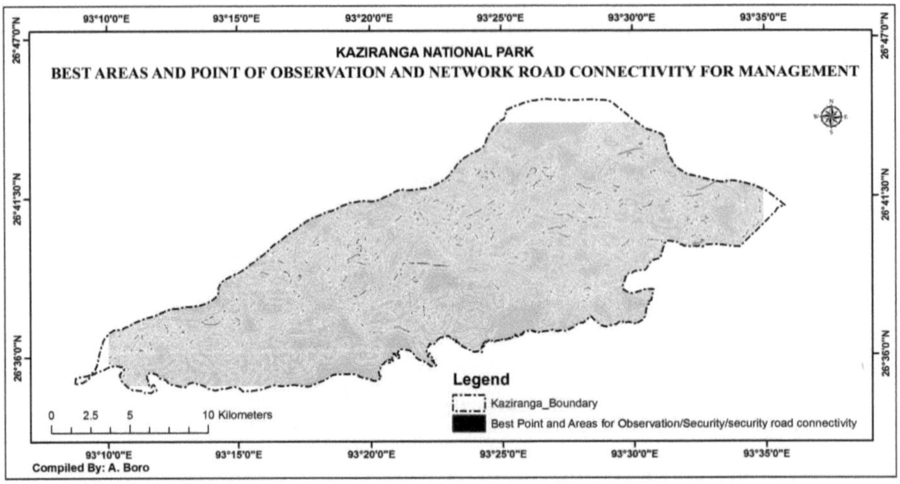

Fig. 16.5 Generated best areas and points of observation and construction of security connectivity roads for management and monitoring

The above three layers on map are generated for the identification of best points and areas of observations which are estimated as requisite for security measures and construction of security road. It is also estimated that connecting points and areas of all the bordering areas of three layers are best locations to be found out. At the same time the bordering areas of wetlands which are 100–150 m away from the wetland margin are considered here as best location for wetland observation. Again, the forest and vegetation layers are allotted 40 % weightage each but the wetland layer is allotted 20 % weightage considering less risk of wetland - concentrated fauna. These conditions can be changed with ground test for practical application.

The selected layers are generated and converted into raster layers in GIS environment. Using the raster calculator and putting the considered conditions in Arc GIS the required information layer is generated (Fig. 16.5). Thus more and more complicated geospatial problems including management of Bio-diversity can be effectively done using this tool and technique.

16.9 Conclusion

The rich Bio-Diversity is an essential and core part for the existence of ecosystem of any area or region for the long run ahead. Management and conservation of it is realized seriously by human society now. On the other hand increasing population in some parts of the world and their activities for survival sometime knowingly or unknowingly use to encroach and destroy biodiversity and environment. Diffusion of awareness to all people and measures for protection has become, therefore, important issues. The technical field of measures for management of Bio-Diversity can be effectively implied in this respect with the help of Geoinformatics. The rich Bio-Diversity of North East India should take advantage of these concepts and tools for the better management and conservation of the region's natural resource for longer possible period of time. So is the case with the Kaziranga NP.

References

Bonham-Carter GF (1994) Geographic information systems for geoscientits: modelling with GIS. Peramon, Love Printing Service Ltd, Ontorio

Booth B (2000) Using ArcGIS 3D analyst. Redlands, Environmental Systems Research Institute, Inc, Relands, CA 92373-8100, USA

Government of Assam (2009) Statistical hand book. Directorate of Economics and Statistics, Guwahati

Hegde SN (2000) Conservation of North East Flora, Arunachal Forest News, State Forest Research Institute, Department of Environment and Forests, Government of Arunachal Pradesh 18(1&2):5–26. http://www.sfri.org/publication.htm

Heywood I, Cornelius S, Carver S (2003) An introduction to geographical information systems. Pearson Education, Delhi

Johnston K, McCoy J (2001) Using ArcGIS spatial analyst. Environmental Systems Research Institute, Inc, 380 New York Street, Relands, CA 92373-8100, USA

Lillesand TM, Keifer RW (2004) Remote sensing and image interpretation. Wiley, New York

Myers N, Mittermier RA, Mittermier CG, da Fonseca GAB, Kent J (2000) Biodiveristyhotspots for conservation priorities. Nature 403:853–858

State of Forest Report (2003) Forest survey of India, Ministry of Environment and Forests, Government of India

UNESCO – IUCN Enhancing Our Heritage Project (2003) Monitoring and managing for success in natural world heritage sites initial management effectiveness evaluation report. Kaziranga National Park, Assam

Chapter 17
Changing Habitat and Elephant Migration from Dalma Wildlife Sanctuary, Jharkhand to Panchet Forest Division, Bankura, West Bengal: A Biogeographical Analysis

Nilanjana Das Chatterjee and Soumendu Chatterjee

Abstract Animal migration involves two aspects- the behavioural characteristics of the animal species and the ecological carrying capacity where it colonizes. Anthropogenic perturbations in the form of agriculture and settlement expansion result in depletion or loss of natural habitat, forest corridors and biodiversity. Disproportionate sharing of forest resources between man and animal causes shrinkage of the sojourn ground and generates food scarcity for wild animal which ultimately forces animals to be evicted from their home habitat and migrate to a new habitat. In course of this migration man-animal conflict becomes inevitable in the forest corridors and edges. Such a typical situation of migration event and man-elephant conflict has arisen in the Panchet Forest Division of Bankura District, West Bengal since 1980s. This paper aims to assess the ecological characteristics of both the source region (Dalma Wildlife Sanctuary, Jharkhand) from where the elephants are forced to migrate and the destination region (Panchet Forest Division, West Bengal) which attracts the elephants to invade. Multi-date satellite images have been analyzed and ecological surveys at selected sites have been carried out to trace changes in ecological characteristics of both Panchet and Dalma areas. Secondary data on damages due to elephant invasion and number of elephants in a herd have been collected from respective Government departments. Field survey has been conducted to understand nature and pattern of elephant migration and its changes over the years. The study reveals that number of migrating as well as

N.D. Chatterjee (✉)
Department of Geography and Environment Management, Vidyasagar University, Midnapore 721102, India
e-mail: nilanjana_vu@mail.vidyasagar.ac.in

S. Chatterjee
Department of Geography, Presidency University, Kolkata 700073, India
e-mail: scgeovu@yahoo.co.in

M. Singh et al. (eds.), *Climate Change and Biodiversity: Proceedings of IGU Rohtak Conference, Vol. 1*, Advances in Geographical and Environmental Sciences,
DOI 10.1007/978-4-431-54838-6_17, © Springer Japan 2014

residential elephants have significantly increased and so is the trend of damage volume. Change of landuse along forest margin tracts has been instrumental in migration-route shifting. Outcomes of this study may help to identify the causes behind elephant migration and formulate management strategies by judging feasibility of management alternatives.

Keywords Ecological carrying capacity • Forage behaviour • Habitat fragmentation • Habitat requirements • Man–animal conflict • Species composition

17.1 Introduction

Habitat selection of an animal species depends on the interactions among behavioural characteristics of the concerned species and the ecological carrying capacity of the habitat where it inhabits. The forage behaviour of mega herbivores like elephants require their range to be extended across kilometers for meeting their fabulous demand of fodder, water, shelter and other biological needs. The phenomena of elephant migration thus depend on the habitat character of where they originally inhabit and the habitat of the new place where they tend to migrate. Elephants, the largest land animals, share their habitat all over the world especially in Africa and Asian Subcontinent. Both in Africa and Asia, there have been numerous studies of the foraging ecology of the elephant. The diversity of habitats in which elephants are found across the two continents, shows variation in the types and parts of plants they consume (Sukumar 1992). Elephants are found to reside in a diverse array of habitats ranging from Rain Forest through Deciduous Forest and Savanna Woodland to near Desert. Hence, their habitat selection mainly depends on the availability of food, water and shelter. They have to migrate from one place to another in search of food. Migration is an adaptation to those resources that fluctuate spatiotemporally either seasonally or less predictably (Dingle and Drake 2007).

In south West Bengal, elephants were abundant in dense *sal* (*Shorea robusta*) forests of Medinipur district and its adjacent areas during early 1900s (O'Mally 1911). Reappearance of elephant in this area started beyond mid 1980s with the revival of forest cover as a result of successful participatory forest management involving local communities (Palit 1991; Malhotra 1995). The forest cover change detection study based on satellite images (1978–1991) confirmed positive increase in forest cover of Bankura, Purulia and Medinipur districts (Sudhakar and Raha 1994). Problem in the study area started with the intrusion of elephant from Dalma Wildlife Sanctuary, Jharkhand. Movement of the elephants deepened towards east in the consecutive years (Chowdhury et al. 1997; Datye and Bhagawat 1995 cited in Singh et al. 2002). Close encounters between human and elephant in this landscape with settlements and croplands, have detrimental effects. For proper management of the situation, understanding the relationship between landscape elements such as land use pattern, habitat assessment, habitat remnants and elephant activity is an important area of research (Kumar et al. 2010). From the forest department's reports

(DFO, Bankura), it appears that there is an increasing trend in the number of migratory elephant into the study area. In 1987, a herd of 40 elephants entered into the study area which increased to72 by 2007. The duration of their stay in the area has also increased with the passage of time. During 1995–1996, the number of days the elephants stayed in Panchet Forest Division was 31 which increased to 70 days by 2006–2007. Damage profile (human injury and death, property loss and damage of agricultural crops) also accrued over the years.

Practicable management strategies to combat the situation can be made out only if they are based on bio-geographical understanding of the migration process that leads the situation to become conflicting. In the processes of sharing resources, human activities result many consequences in the form of decline in quality and coverage of forests, loss of biodiversity, removal of forest corridors and conversion of land which ultimately configure the degraded habitat of the wild species. Such anthropogenic changes may restrict or modify the tracks of animal movement. The above mentioned factors squeeze elephant herds in protected forest areas and their ranging pattern which are causing forceful changing of habitats by the elephants. The study seeks to identify the changes in structural characteristics of forest areas under study and their bearing upon increasing trend of elephant migration, if any, and also to identify the temporal pattern and damaging effects of elephant migration on society.

17.2 The Study Area

17.2.1 Dalma Wildlife Sanctuary, Jharkhand (Source Area)

It is situated in the Purbi Singhbhum District of Jharkhand State (between 22°46′30″N & 22°57′00″N latitudes and 86°03′15″E & 86°26′30″E longitudes) and has an area of 193.22 km^2 (Fig. 17.1). The area is characterized by undulating terrain with high hillocks (highest elevation 984 m above MSL), plateau, deep valley and open fields between hillocks, providing diverse habitat for flora and fauna. The forests here are mostly of Dry Mixed Deciduous type dominated by Dry Peninsular *Sal*, the main tree species being *Terminalias, Jamun, Dhaura, Kendu, Karam* etc. The sanctuary is very much favoured by the elephants due to availability of water even during the summer months. Dalma Wildlife Sanctuary (DWS, declared sanctuary in 1976) is situated in the Purbi Singhbhum and Saraikella Kharswan districts of Jharkhand. DWS spreads over 193.22 km^2 area (of which 35 km^2 belongs to core area and 158.22 km^2 to buffer area) and constitutes a part of R. Subernarekha catchment. The Sanctuary is surrounded by 85 villages. This sanctuary is declared as a natural habitat for elephants.

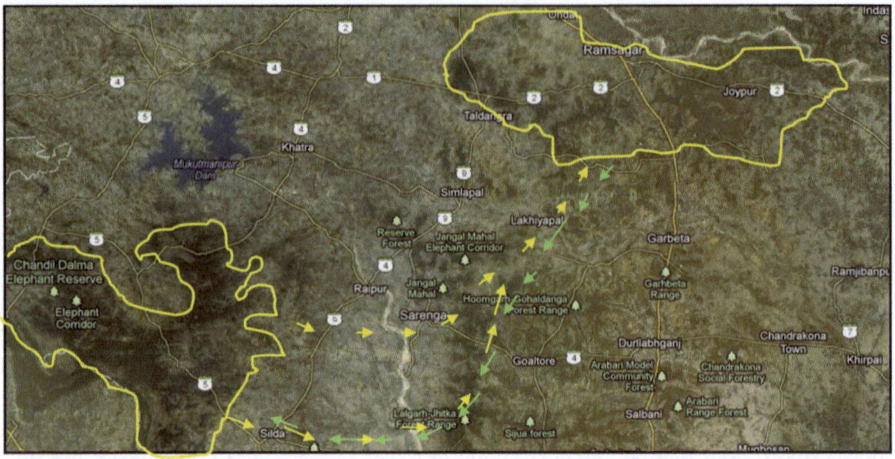

Fig. 17.1 The Source and Destination areas in open source satellite image. Out migration route (*Green right arrow*). In migration route (*Yellow right arrow*)

17.2.2 Panchet Forest Division, Bankura,West Bengal (Destination Area)

Geographically it extends between 22°53′N & 23°12′00″N and 87°03′00″E & 87°42′00″E. Panchet Forest Division covers an area of 336.04 km^2 distributed among 7C.D Blocks of Bankura District,West Bengal (Fig. 17.1). The forest area is divided into 5 territorial forest ranges which are divided into a total of 21 forest beats which include 236 forest bearing mouzas. It is situated in the eastern fringe of Chhotonagpur plateau.

Both the areas are characterize by *patches* represented by forest remnants; isolated and degraded pockets of the forest; *corridors,* represented by rivers, railway tracks and roads that fragment the habitat and *background matrices* represented by continuous stretches of forests or agricultural lands.

17.3 Data and Methods

The study is an intensive field based empirical research using both qualitative and quantitative data and methods. The work seeks to verify the ecological principles of animal migration and aims to explain the causes behind elephant migration adopting direct observation method. It is tried to explore the trend or temporal pattern of elephant migration and its relation to gradual deterioration of forest habitats over the years. Emphasis has been given on characterization of the man elephant conflict zones i.e. the forest margin villages or destination areas of the migration events that exert pulling forces to attract the elephant (Barua and Bist 1995).

Attention has been paid on assessment of the changing landuse- landcover characteristics through ecological monitoring. Habitat quality has been evaluated through assessment of biodiversity in terms of species richness, species density, species abundance, and composition and association of plant species. Cropping pattern of the destination region has been considered as a factor of attracting the elephants. Moreover, huge loss of agricultural products is suffered by the farmers every year. A vivid field enquiry has been carried out to address damage issues. Ecological field survey techniques have been adopted along transacts and across quadrats which were randomly selected from the entire field area. Herbarium were prepared for each sample field site and they were identified for assessment of the quality of habitats of both source and destination regions. Landscape analysis is carried out using LISS-IV + P6 merged images of 1980, 1990, 2000 (at an interval of 10 years). Ecological characters of forests like association and density of plant (trees, shrubs and herbs) species, forest density, canopy cover, foliage cover, pattern of fragmentation of each of the forest types in the study area (as identified from image analysis) are enumerated. Year wise secondary data and information on the number of migrated elephants and migration routes have been collected and analyzed. Anthropogenic encroachment in forest and change in landuse/landcover pattern is evaluated using digital images.

17.4 Results

17.4.1 Habitat Characteristics

17.4.1.1 Dalma Wildlife Sanctuary

DWL is located in the Purbi Singbhum district of Jharkhand state. ‘Jharkhand’ or ‘the land of forest’ is situated within Chhotonagpur plateau region. The state is endowed with mineral deposits and thick forest cover. The region has huge reserve of coal, iron ore, mica, bauxite and limestone and considerable reserves of copper, chromites, asbestos, kyanite, china clay, manganese, dolomite, uranium etc. (Fig. 17.2). Most of the mining areas are situated in the Purbi Singbhum district (Fig. 17.5). Huge extractions of minerals and rapid industrialization in this state, particularly in the Purbi Singbhum district have caused large scale destruction of forest in DWS.

According to the Forest Report published in 2011, Purbi Singbhum district had 30.51 % forest cover to its total geographical area (Fig. 17.3). Between 1997 and 1999, about 3,200 ha of forest cover was lost in the Singhbhum region. Between 2001 and 2003 some 7,900 ha of dense forests were lost in the Purbi and Paschim Singhbhum districts. The main reason behind this is uncontrolled mining for iron ore, both legal and illegal. Mining is destroying not just the forest, but also the wildlife, apart from the livelihoods of the local tribal communities. Sal (*Shorea robusta*) is the dominant tree species in DWS area (Table 17.1 and Fig. 17.4).

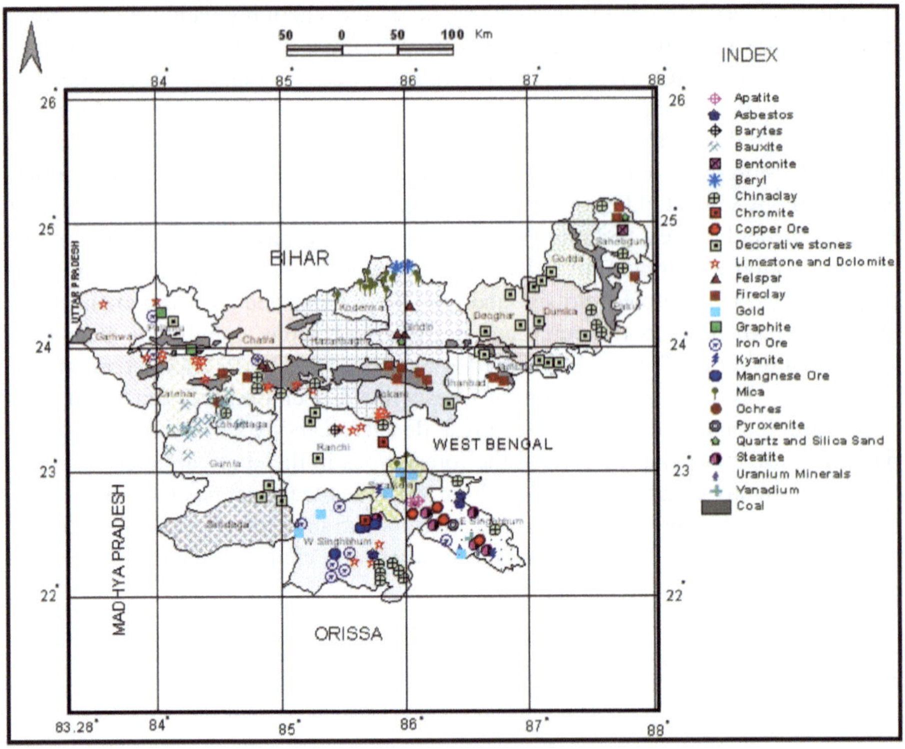

Fig. 17.2 Distribution of mineral resources in Purbi Singhbhum District, Jharkhand (Source: NRDMS, Jharkhand)

Species diversity is higher in the Dalma forest region than the Panchet Forest Division. Varieties of trees, shrubs, herbs and climbers are found. But due to large scale mining and allied activities forest areas get severely damaged. Extension of agricultural land into the forest areas is also responsible for fragmentation in the edge areas. Moreover, the sanctuary area is surrounded by settlements and mining spots. Blasting in the open cast mining areas disturbs the calm environment of the sanctuary. According to census report of Jharkhand (1991) during 1980s coal companies acquired thousands of hectares of forest in Jharkhand for mining operation in Damodar valley. The dense natural vegetation cover has gone down to only 29 % of its total area which not only reduced the biodiversity but also minimized the habitat area for wild animals.

17.4.1.2 Panchet Forest Division

Natural vegetation of this area has been strongly guided by rainfall, temperature and soil. During the last twenty five years the nature of forest cover of the area has significantly changed. It is seen that after the implementation of Joint Forest Management, there has been an increase in the forest green cover. Much of the

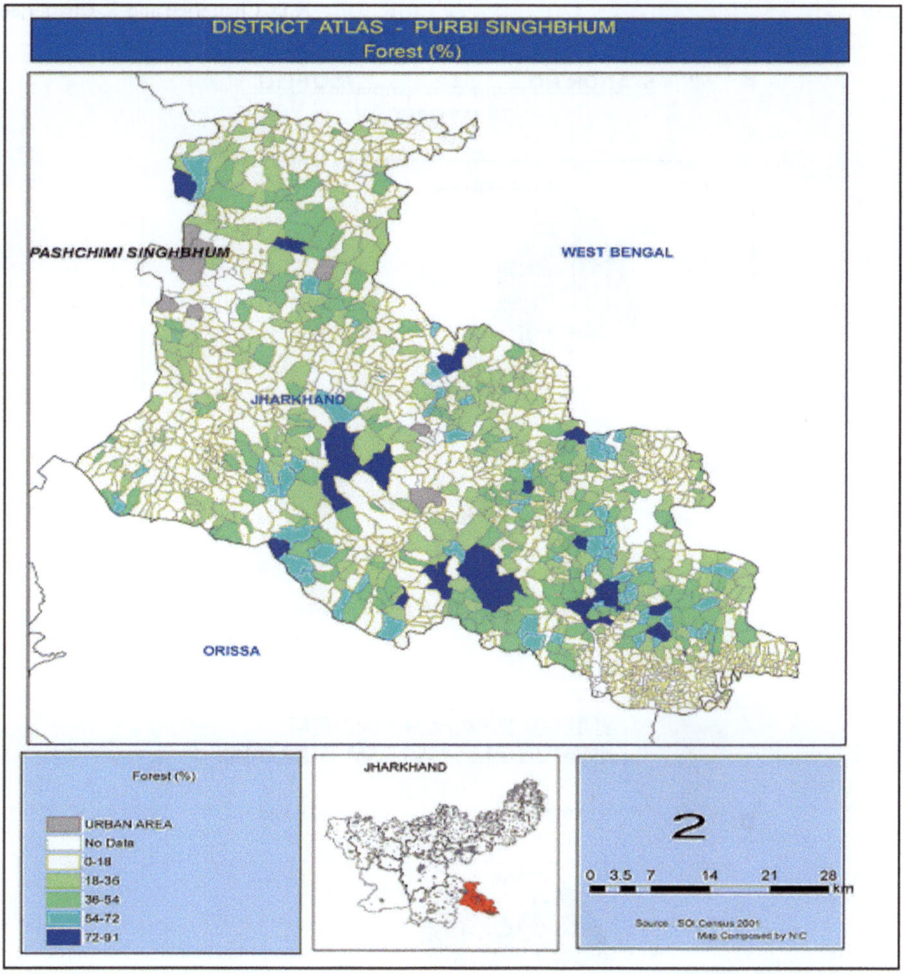

Fig. 17.3 Forest Density of Purbi Singhbhum District, Jharkhand. *Source*: NRDMS, Jharkhand

Table 17.1 Dominant plant species in Dalma forest

Local name	Botanical name	Local name	Botanical name
Babul	*Acacia arabica*	Kathul	*Bauhinia retusa*
Dhela	*Alangium Lamarckii*	Salia	*Boswellia serrata*
Siris	*Albizzia lebbek*	Kajhi	*Bridelia retusa*
Jang Siris	*Albizzia odoratissima*	Piar	*Buchanania lanzan*
Safed Siris	*Albizzia procera*	Palas	*Butea frondosa*
Chatni	*Alstonia scholaris*	Kumbhi	*Careya arborea*
Dhautha	*Anogeissus latifolia*	Beri	*Casearia tomentosa*
Bhabiranj	*Antidesma ghaesembilla*	Dhanraj/Amaltas	*Cassia fistula*
Kathal	*Arotocarpus integrifolia*	Bharhul	*Chloroxylon swietenia*
Barhar	*Artocarpus lakoocha*	Belwanjan	*Cordia Macleodii*
Neem	*Azadirachta indica*	Bahuar	*Cordia myxa*

Data compiled through grid-based ecological survey at sample sites in Dalma forest area

a

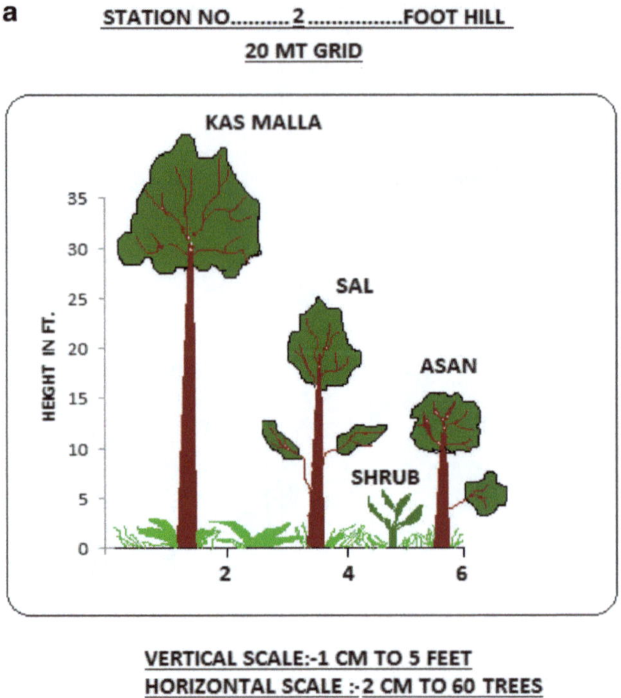

b

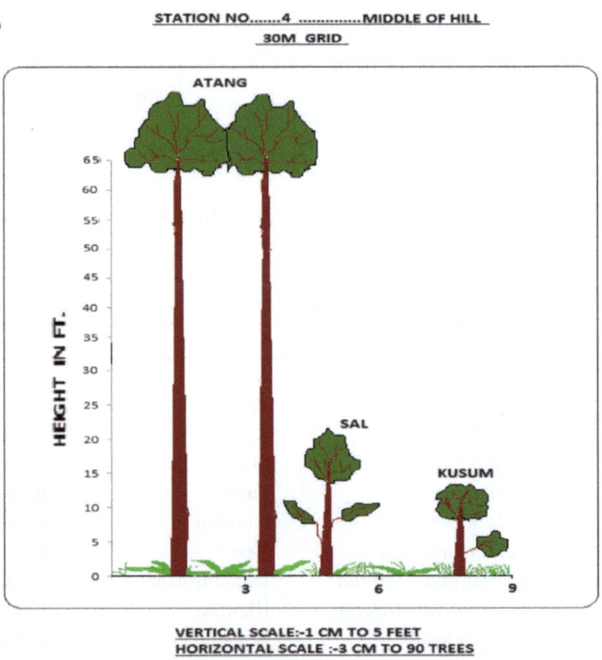

Fig. 17.4 Schematic diagrams to represent plant species composition and forest structure at sample sites in Dalma Wildlife Sanctuary, Jharkhand (**a**) Foothill site, (**b**) Middle slope site along forested hill

Table 17.2 Major plant species (trees, shrubs and herbs) in Panchet Forest Division, W.B

Local name	Botanical name	Local name	Botanical name	Local name	Botanical name
Am	*Magnifera indica*	Akanda	*Calotropic gigantia*	Kulekhara	*Asteracanlta longifolia*
Arjun	*Terminilia arjuna*	Tulsi	*Asimum sanctum*	Thankuni	*Centella asiatica*
Asan	*Terminalia tomentosa*	Dhutura	*Datura stramonium*	Ghritakumari	*Aloe vera*
Cashew	*Anacardium occidentalo*	Kul	*Zizyphus xylopyra*	Harjora	*Vitis quadrangularis*
Kusum	*Schleichera trijuga*	Nishinda	*Vitex negundo*	Kalmegh	*Andrographic pariculata*
Mahua	*Madhuka latifilis*	Lajjabati	*Minosa pubica*	Kurchi	*Hylarhena antidyshterica*
Minjiri	*Cassia siamea*	Sialkanta	*Argemone maxicana*	Susni	*Marsilea miiluta*
Neem	*Azadirachta indica*	Satamuli	*Sparagus racemosus*	Basak	*Adhatoda basika*
Palash	*Butea monosparma*	Alkushi	*Mucuna prurita*	Haritaki	*Terminalia chebula*
Simul	*Bombax malabaricum*	Chhatim	*Alstonia scholaris*	Bahera	*Terminalia balarica*
Siris	*Albizzia lebbek*				
Teak	*Tectona grandis*				

barren refractive soils have been turned into green with Akasmoni (*Accacia auriculiformis*) and Eucaliptus. A notable change in the floral diversity is found in this part of the study area (Table 17.2). It creates natural corridor for elephants. A forest density map for this area has been generated from the data received from the Forest Department (Fig. 17.6). Maximum density is found in the western part of the study area and it decreases towards the east due to expansion of agricultural lands.

In Panchet forest region the patches are dominantly composed of degraded lands where forest has been cleared in isolated pockets. The linear remnants and regenerated strips of forest and valley fills are the important corridors for the migration of elephants. The entire forest tracts and agricultural fields constitute the back ground ecological matrices. Lands of the area have been given under various uses. River Darakeswar flows from the west to the east of the study area. The area is traversed by roads and railways. Several metalled, unmetalled, rural roads and cart track have dissected the entire area (Fig. 17.1). Settlements are distributed all over the study area but they are more numerous in the eastern part due to availability of fertile alluvial tracts.

Structurally the forest is characterized by a tree layer, a shrub layer and a herb layer (Table 17.1 and Fig. 17.5). Forest habitats are more fragmented along their margins due to conversion of forest lands into agricultural lands. As a consequence, the continuity of larger habitats which is broken which is a threat to the natural ecosystem (Fig. 17.6). Deforestation in the marginal areas is also responsible for fragmentation of habitats (Knickerbocker and waithaka 2005).

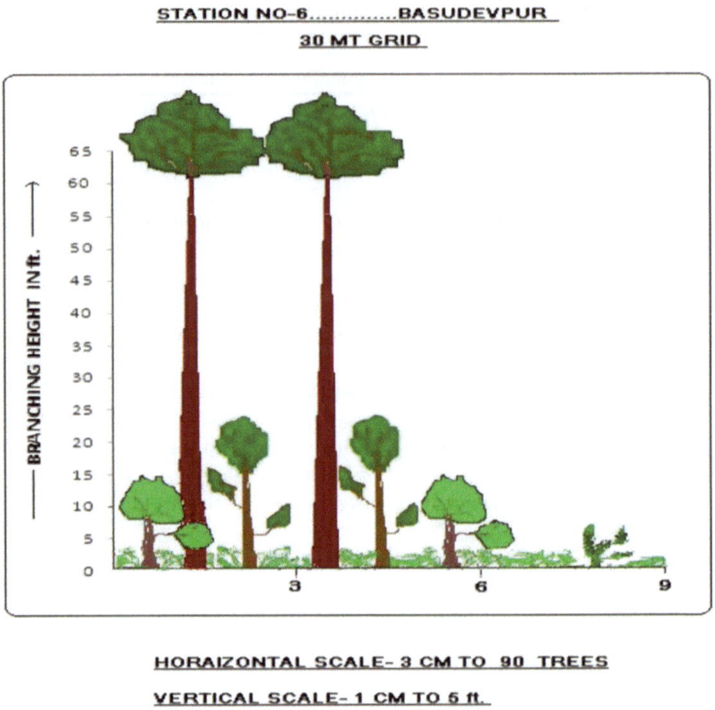

Fig. 17.5 Schematic diagram representing plant species composition and forest structure at a sample site in Panchet Forest Division

Fig. 17.6 Forest density in different beats of Panchet Forest Division, W.B

Both Dalma and Panchet forest areas share certain common characteristics which are important so far as elephant habitat is concerned.

- Seasonal rainfall, periodicity in plant growth and reproduction.
- For vegetal feeding animals food supply is less varied and less continuously available.
- In certain places and in certain time foods become more abundant.
- Foods are easily accessible because forest is less dense.
- Patches of cultivated lands within or at the margins of forests acts as centre of attraction as edible substances are concentrated here.

17.4.2 Shifting of Migration Route

From the forest report it is found that since 1987 the event of elephant migration from Dalma forest to Panchet forest had become regular phenomena. In that year a herd of 40 wild elephants moved towards east of Dalma. They crossed Kansabati river and entered Lalgarh range area of Paschim Medinipur Division. Since 1988 they invaded into Bishnupur forests after crossing Silabati River and since 1995 they have been crossing Darakeswar and moving towards northern forest division of Bankura District. They extended their territory up to Beliatore forest since 1999. During this period another new herd crossed Damodar River to enter Bardhaman District. There are about 3–4 herd each consisting of 12–50 elephants used to come each year in the Panchet Division. In 2012, there were more than 180 elephants invaded Panchet forest region. They moved around Bankadaha, Joypur and Bishnupur Range. So there is a permanent elephant migration route from Dalma to Panchet and adjoining areas. Along with the increasing number of elephants in the destination region the duration of stay has also increased. During 1995–1996, the number of days the elephants stayed in Panchet Forest Division was only 31. In 2006–2007 this number increased to 70 days and in 2012 it further increased to 180 days. As the duration of stay increased, man-animal conflict became more frequent and volume of destruction of agricultural crops has also increased.

During 2003–2004, 4,680 people were reportedly affected which was 1,357 in 1995–1996; and during 2004–2005 crops of 75 ha of lands were damaged while this figure rose to 172.01 ha by 2007–2008. The above statistics clearly demonstrate enormity and severity of the problem. Examination of the temporal shift of migration routes reveals that they are entering the Panchet area by same route but their return route is shifting towards the east where the land under agriculture is gradually increasing (Fig. 17.7). Continuously they are extending their habitat by adopting new strategies. They have changed their food habits from agricultural crops to juicy and palatable horticultural crops like cabbage-cauliflower, cucurbits, pumpkin, potato, brinjal, jack fruit and ketsu etc. Finally, it becomes evident that the Panchet Forest habitat, which was once the temporary habitat of the Dalma elephants, is gradually becoming the permanent habitat of these elephants.

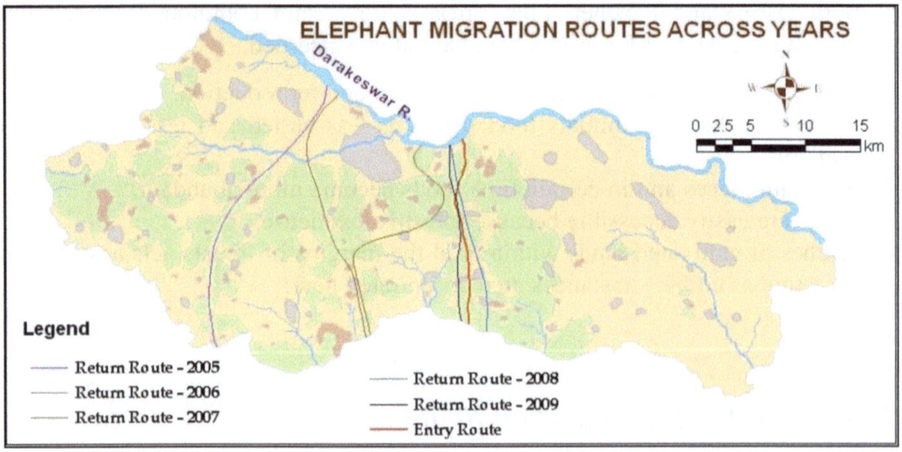

Fig. 17.7 Route taken by elephants in recent past years while returning to the source region

17.5 Discussion

This study reveals that the event of elephant migration from DWS to Panchet Forest Division has become a issue of great concern. Both the number of migrated elephant and duration of stay in the new habitat are increasing. Increasing elephant migration may be attributed largely to disturbances created during mining-quarrying activities, mine blasting and construction of industries at the cost of habitat peace in Dalma area. Degradation and fragmentation of habitat by anthropogenic activity are also responsible for elephant migration. Though the forest density and species diversity of DWS is more than Panchet Forest area but Panchet area is less disturbed (there is no large scale industry or mineral resource mining centre). Our survey on the habitat preference of elephant reveals the fact that they prefer those areas which are less intervened by humans along with the availability of two other food and water. Forest cover of Panchet area is fragmented too but is covered by secondary vegetation which is a favourite diet for the elephants (Sukumar 1990). The forage behavior of the elephants demands following habitat requirements which are largely met in Panchet forest area.

- Elephants are wide ranging animals requiring larger extent of continuous stretches of forests for food, shelter and water (Fernando et al. 2008)
- A herd of 100 elephants would require a minimum of about 650 km^2 of area.
- Home range size varies according to topography and types of vegetation.
- Spend about 70 % of the time for feeding which varies with season.
- Grasses, bamboo, paddy, any kind of vegetation are the main foods of the elephants.
- Iron, Copper, Boron, Calcium and Sodium are the important minerals required for which they have developed bark feeding behaviour.
- Consume 1.5 % of their body weight in 12 h feeding.
- Require 100 L of water at one time and 225 L of water in a day.

17.6 Conclusion

Since 1987, large numbers of elephants from Dalma Hills of Purbi Singbhum district of Jharkhand, have been entering the densely populated districts of Midnapore and Bankura in West Bengal (Wildlife Protection Society of India). This fact matches with the incidence of large scale degradation of forest due to mining activities and industrial development in Dalma Wildlife Sanctuary during this period. Another important event in this context is successful accomplishment of Joint Forest Management programme in West Bengal. A vast area of degraded *sal* forest had been turned into newly generated forest in Panchet area that has provided convenient shelter and new corridors for migrating elephants. The fragmented forest patches are surrounded by villages and agricultural fields with ample paddy, vegetable and water. All these resources attract migratory elephant. Since 1987 they had been invading Panchet forest regularly. Our field enquiry on the habitat preference by elephants shows that the area having thick ground cover, isolated forest patches near the settled areas and availability of water sources are the most preferable place for elephants (Fig. 17.8).

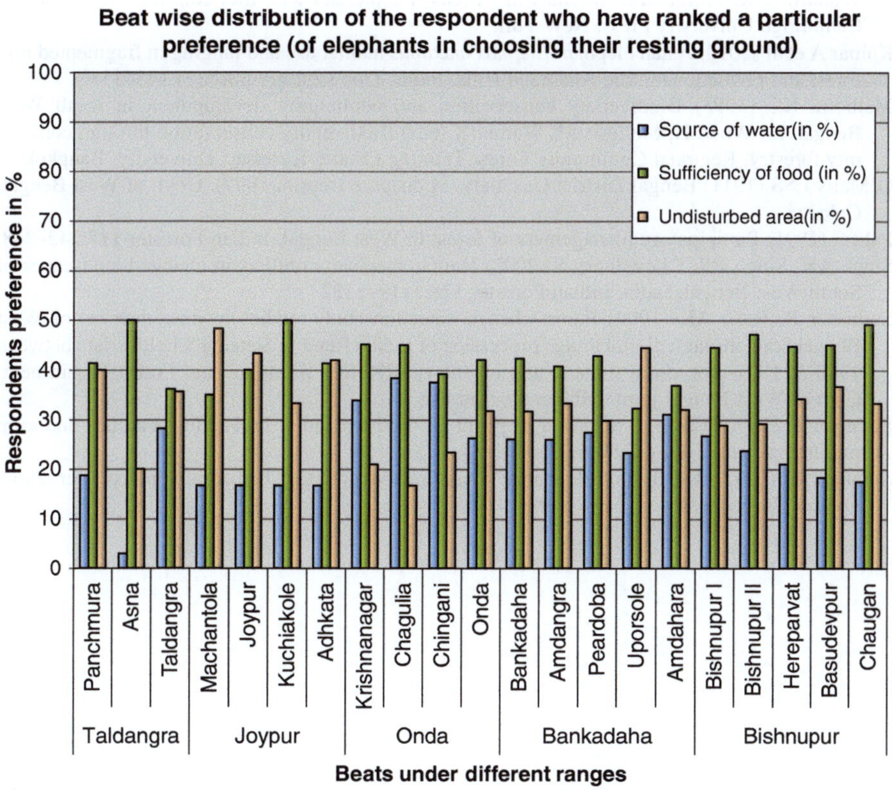

Fig. 17.8 People's perception about characteristics of resting ground preferred by elephants

As such, the Dalma elephants have developed a habit over the years to migrate to the villages in fringe areas of Panchet Forest Division being attracted by paddy and other crops and vegetables cultivation of which has dramatically increased in those areas during last few decades.

References

Barua P, Bist SS (1995) Changing patterns in the distribution and movement of wild elephants in North Bengal. In: A week with elephants, BNHS, Bombay

Chowdhury S, Khalid MA, Roy M, Singh AK, Singh RR (1997) Management of elephant population in West Bengal for mitigating elephant-human conflict, Final report submitted to West Bengal Forest Department. Wildlife Institute of India, Dehra Dun, India

Datye HS, Bhagawat AM (1995) Home ranges of elephants in fragmented habitats of central India. J BNHS 92:1–10

Fernando P, Kumar MA, Williams AC, Wikramanayake E (2008) Review of human elephant conflict: mitigation measures practiced in South Asia (AREAS technical support document submitted to World Bank), World Bank-WWF Alliance For Forest Conservation & Sustainable Use

Dingle, Drake (2007) What is Migration? Bio Sci Mag 57(2):113

Knickerbocker TJ, Waithaka J (2005) People and elephants in Shimba Hills, Kenya. In: Woodroffe R, Thirgood S, Rabinowitz A (eds) People and wild life: conflict or coexistence. Cambridge University Press, New York

Kumar A et al (2010) Asian elephant *Elephas maximus* habitat use and ranging in fragmented rain forest and plantations in the Anamalai Hills, India. Trop Conserv Sci 3(2):143–158

Malhotra KC (1995) Biodiversity conservation and community development in south West Bengal. In: Wood H, Mellises M, Warner K (eds) Biodiversity conservation through community forestry. Regional Community Forest Training Centre, Kesetsart University, Bangkok

O'Mally LSS (1911) Bengal District Gazetters, Midnapore (reprint 1997). Govt. of West Bengal, Calcutta

Palit S (1991) Participatory management of forest in West Bengal. Indian Forester 117:342–348

Singh AK, Singh RR, Chowdhury S (2002) Human elephant conflicts in changed landscapes of South West Bengal, India. Indian Forester 128:1119–1132

Sudhakar R, Raha AK (1994) Forest change detection study studies of nine districts of West Bengal State through digital image processing of Indian Remote Sensing Satellite data between 1988 & 1991-procedural manual and inventory. RRSSC, Khragpur and Forest Department, Govt of West Bengal joint collaborative project

Sukumar R (1990) Ecology of Asian elephant in southern India. II. Feeding habitats and crop raiding patterns. J Trop Ecol 6:33–53

Sukumar R (1992) The living elephants: evolutionary, ecology, behaviour, and conservation. Oxford University Press, New York, p 191

Chapter 18
What Can Human Geography Offer Climate Change Modelling?

Thomas Skou Grindsted

Abstract The discipline of Geography may be one of the most prominent and oldest disciplines in the conceptualization of human–environment interactions that integrates elements from both natural and social sciences. Yet, much research on society–environment interactions on climate change reduces human behaviour to economic rationality when construed in sophisticated climate models and sometimes in non-geographical representations. The need to comprehensively take into consideration methodological approaches concerning the interface of society-environment interactions seems highly relevant to contemporary conceptual modelling of climate change adaption and mitigation. In other words, geographical representations do matter. In the following we will first reflect upon what I shall call spatio-temporal tides and waves of the human environment theme to examine the methodological grounds on which climate change models is based. From a history-geographical perspective the article shows that notions of objective models are increasingly challenged in an era of the anthropocene. It points toward a discussion of interdisciplinary challenges and the ways in which different traditions interpret and explain regularities, rationalities, and pre-analytic assumptions. Lastly we discuss challenges of constructing nature(s) and how we better understand the (geo) politics of climate change modeling.

Keywords Climate change • Conceptual models • Human–environment • Second nature • Geo-politics of modelling

T.S. Grindsted (✉)
Department of Environmental, Social and Spatial Change, Roskilde University, Roskilde, Denmark
e-mail: tskoug@ruc.dk

M. Singh et al. (eds.), *Climate Change and Biodiversity: Proceedings of IGU Rohtak Conference, Vol. 1*, Advances in Geographical and Environmental Sciences, DOI 10.1007/978-4-431-54838-6_18, © Springer Japan 2014

18.1 Human Environment Interfaces in an Era of the Anthropocene

In general consensus exists among scientific and geographical communities as to the cause and effect of climate change and unsustainable production patterns. Consensus exists to the extent that 97 % of research articles in high-impact factor journals like Science suggest that climate change is fundamentally anthropogenic (Oreskes 2004). Correspondingly, Cook et al. (2013) find 97.1 % of more than 4,000 peer reviewed articles from the past 20 years examined support that global warming is mainly or entirely human induced (Cook et al. 2013). Today, man interacts with the physical environment to the extent that humans are transforming the planet from one geological epoch (the Holocene) toward a new geological era; the anthropocene (Crutzen 2002). The anthropocene refers to the magnitude, scale and acceleration of per capita exploitation of natural resources that transform the biogeography for millennia to come. The history of human–environmental interaction is indeed astonishing. During the past century we have witnessed massive land cover and land use changes of the Earth. From 1900 to 2011 the world's population has grown by a factor four (from 1.6 to 7 billion in 2011) accompanied by a growth in cattle and pig production to more than 1,400 and 800 million respectively.[1] Irrigated areas are five times the amount. Urbanization grew by a factor of 13, energy use by a factor of 16, and industrial output by a factor of 40 (McNeill 2000; Crutzen 2002; Steffen et al. 2011; Hertel 2011). In the same period rainforests were reduced 20 %. Deforestation, however, is only a droplet compared to the reduction of manifold biotopes by the agricultural demand for cropland. Today, more than half of the world's land surface has been changed by human activities which illustrate the very need for geographical representations in understanding transformations of the Earth life support system (Reenberg 2006; Griggs et al. 2013).

The journey of geographical transformations is also a journey of the nature of time and space (Massey 1999). The changing geography of the world's physical environment, the biogeography and land use mutually transforms humans and their environments. Therefore methodological and geographical reflections of the human environment interactions seem more relevant than ever. Prediction of future climates and planetary constrains are indeed beneficial and the geographical imagination is central to conceptual climate and land use model building (O'Sullivan 2005; Mendelsohn and Dinar 2009). By way of illustration, Eugene Linden showed how interdisciplinary constructs coupled spatial imaginations through satellite images needed to be assembled before a unified account of past, present and future climate data formed collections of global assemblages of explaining the

[1] The growth in cattle and pig production from 1900 to 2010 grew by a factor of 4 and 9 respectively.

climate systems: “A system in which everything, from earth’s position in its orbit around the sun to what’s growing on the ground, influences climate. How the climate system balances these various inputs and feedbacks is a problem complex as life itself” (Eugene Linden here quoted in Urry 2011, p. 23).

18.1.1 Man and Nature: Integration of Data and Disciplines

Human and physical geography will change remarkably through decades to come if the processes of climate change predicted is even half right (IPCC 2013). Global Climate Models (GCM) integrate Earth Observation Data (EO), Remote Sensing (e.g. Landsat) coupled with socio-economic data that help us understand the material and biogeographically transformation of the environment (Reenberg 2006; Dangermond and Artz 2010). The study of human–environment interfaces, however, is a subject in which many traditional disciplinary often fail to properly address methodological, epistemological and ontological pre-analytic assumptions in time and scale (Rasmussen and Arler 2010). In fact many contemporary challenges cannot be adequately addressed within the boundaries of traditional disciplines. “Even that ingrained counterposition between so-called ‘natural’ and ‘social’ is increasingly being questioned, and my conviction is that if they are now up for reinspection and problematization, then geographers should be in a good position to make a leading contribution” (Massey 1999, p. 261).

The idea of institutionalizing interdisciplinary approaches constantly challenge traditional disciplinary boundaries of human–environment interface(s), e.g. Human Ecology, Environmental Studies, Earth System Science, Geography, Ecological Economics, Landscape Ecology and Sustainability Science. There is no single “climate science”, but a myriad of overlapping scientific (inter)disciplines with much rivalry and competition (Urry 2011). Divergence and convergence between these contested disciplinary constructs in reorganizing sciences engaged in environmental change, is confronted with a number of multi-scalar methodological problems not to mention constructions of geographical imagination.

By way of example, much contemporary climate change modeling assumes seemingly neutrality and objectivity while at the same time often designed with non-spatial representations (Globium is an exception of the latter). Climate models as well as land modeling are based upon huge amounts of sophisticated statistical properties including assumptions of behavior of many features (human or non-human). Compounded as ‘neutral landscape models’ (Turner 2005, p. 324), these models are (whether GTAP, IMAGE, AgLU, IMPACT, GLOBIOM, ABLUM, GIS or GCM) organized reductions of geographical representations, or more challengingly; super artificial objective reductionism of human–environment interactions often construed as partial or general equilibriums (Hertel et al. 2010).

18.1.2 Anthropogeography

Geographers have long challenged the idea of objective non-human nature, giving rise to concepts like 'second nature', the 'politics of Nature' or even 'multinatural ontologies' well before Paul Crutzen (2002) coined the term 'anthropocene' (Latour 1993; Harvey 1996; Braun 2006; Lorimer 2012).[2] Interaction between the natural and social worlds has indeed proven difficult to conceive epistemologically, e.g. in social physics, when ecological economists seek to integrate the language of biology into economic theory, or more notably when biological concepts have entered social theory in forms of Social Darwinism, Fascism or Nazism (Harvey 1996). Nonetheless, anthropogenic climate change is a socio-material phenomenon and we need better epistemological and methodological approaches to grasp these challenges (Lorimer 2012). Thus, we examine if the multiple traditions of human–environment interactions within human geography have anything to offer climate change modeling, and if so; how it might help us find ways of thinking conceptual climate change models. Can we possible draw some insightful perceptions from the history of human–environmental interactions in understanding the 'nature(s)' of climate change modeling?

18.2 Anthologies and Ontologies of Human Nature(s)

Geography may be one of the most prominent and oldest disciplines in the conceptualization of human-environment interactions that integrates elements from natural and social sciences (Rasmussen and Arler 2010). In fact, "The theme of man-environment relation has never been far from the heart of geographical research, and for many it has functioned as the overriding theme" (Harvey 1969, p. 115). The history of the human environment theme, however, has taken multiple forms and methodological approaches over the years. Some geographers conceptualize the human–environment theme more or less ad hoc, implicitly or explicitly whereas others organize it into constructs separating human and nature or build certain interfaces. Though assumptions of the human environment theme are sometimes implicit they hold 'tacit information' that is mediated through scientific and educational practice (Demeritt 2002). Tacit information neither is logical, consistent nor reflected methodologically, still it carries huge amounts of knowledge that exists in the interface between subject and object, between human and nature relevant to geographical representations of climate change modelling. Therefore we must never ignore the nexus between (tacit) knowledge and power

---

[2] The idea of the anthropocene, can also be traced back to a number of thinkers in the early nineteenth Century, e.g. Valdimir Vernadsky's, mankind's geochemical work, Eduard Suess concept of the anthropogenic transformation of the biosphere into the noösphere or man as an geological agent (Steffen et al. 2011).

e.g. in the construction and use of models. Following Harvey, geographers build explanations on the way a theme is constituted; "A theme acts as a directive by indicating the sort of facts the geographer ought to collect and by suggesting a mode of organization of those facts" (Harvey 1969, p. 116). A theme 'gives rise to theorize' as Harvey puts it, and how the human–environment theme is considered implicitly or explicitly can be examined through the way different kinds of explanations are perpetuated.[3] Dialogue about these issues may not only better prepare students, researchers or agencies dealing with wicked and controversial problems, but may also make us better understand the geopolitics of the 'object of study' that fundamentally shape questions asked and data collected (Braun 2006). I therefore refer to the human–environment theme as organized assumptions about the way we categorize parts of the world that are not only to be considered within the earth system and human system, respectively. Thus, what is to be considered relevant to the human–environment theme varies from discipline to discipline. Whereas the human environment theme organizes the world thematically (Aristotle) human–environment interaction reflects how we build explanation of interactions within the human–environment theme epistemologically (Birkeland 1998). Methodological reflections on relations and dynamics are focal point to explain a given phenomenon and why it is so. Thus, explanation of effects and interaction (either it be causal, intentional, functional or evolutionary) also signify conceptualizations of problems and associated solutions (Hansen and Simonsen 2004).

Smith (2010) formulates 'the production of nature' thesis as a concept that extends spatial theoretical work of 'the production of space' and amalgamates the spatial chronological theme with the human environment theme. For Harvey as well as for Smith, nature and culture are dimensions of the same phenomena continuously knitted, so that "We cannot talk about the world of nature or environment without simultaneously revealing how space and time are being constituted within such processes" (Harvey 1996, p. 263). Massey grasps such methodological reflections wonderfully and shows how nature and society interactions must be studied as "endlessly, mobile, restless, given to violence and unpredictability" (Massey 2006, p. 38). Within such a meta-theoretical framework, epistemology is based on an intra- and extra-discursive reality. This implies that elements independent of human perception are also formed through human practice. At the same time the framework rests on the assumption that habits of thought and societal development and habits of thought and the material world are closely interrelated (Elling 2003). Methodological assumptions concerning determinism, possibilism or descriptive connotations provide radically different answers to such questions and hence different answers to analysing the same phenomena. To overcome such barriers, human–environment interactions can be studied (1) by asking what is related and how, (2) by unfolding concepts and (3) how they are formed and give rise to other

[3] It must be observed, that facts are also a social construction. Facts are not objective, neutral, independent data constellations, but carry the same value-leadenness as does a theme for which reason it is also of importance to find out how facts are socially accomplished.

interpretations (Harvey 1996). The strength of a dialectical framework is that it splinters binary thinking, essentialism, and absolutism of human–environmental conceptualizations (Birkeland 1998). In this way it is possible to comprehend inclusion/exclusion of features and themes as a practice concerning the building of geographical explanation and how it affects analyses, practices or policies (Harvey 1969). From a history- geographical approach we examine human–environment interactions of what I shall call 'spatio-temporal tides and waves' intersecting, overlapping and conflicting. Following Massey (2006), geographical representations are studied as a mosaic of understandings often in opposition to other representations. The history of spatio-temporal tides and waves are examined through the roots of determinism, possibilism, particularism and absolutism and the way different traditions interpret and explain regularities, rationalities and relations. Then, discussed in the context of conceptual models of climate change, we challenge spatio-temporal figurations used in modeling human-nature interfaces. Lastly, we discuss the (geo)politics climate change modeling in an era of the anthropocene.

18.3 Spatio-Temporal Tides and Waves: Co-constructing Nature(s)

Alexander von Humboldt (1769–1859) may be one of the immediate forefathers of modern human–environment interactions in geography and early environmental science (Zimmerer 2006). Humboldt's advocacy of geognocy (todays Earth Science) considerably contributed to modern environmentalism (1846–1862), as well as Joachim Schouw, Vidal de la Blache, Carl O. Sauer, Harlan Barrows among others advocated that geographers should study human beings in relation to their geographical environment (Christiansen 1967; Turner 2002; Zimmerer 2006). As early as 1864 George Pekins March argued in *Man and Nature or Physical Geography as Modified by Human Action* that ancient civilizations collapsed due to environmental degradation. Throughout the 1920s it was suggested that geography effectively were human ecology studying humans influence on and adjustment to the environment. In a historical perspective it is interesting to observe that invitations to upscale ecological themes have been numerous during the past centuries. "The view of geography as human ecology has quite a long history" (Harvey 1969, p. 115) and since Vidal de la Blache or Humboldt geographers like Forsberg (1962), Moss and Morgan (1965) or Stoddart (1965) have argued for up scaling ecological themes in geography, particularly around the concept of system ecology and human ecology (Christiansen 1967). During the past decades, this seems to have been revitalized particularly around agendas of sustainability and climate change due to the study of human–environmental interactions (Zimmerer 2010; Dangermond and Artz 2010). The interface between the spatial chorological approach and the human–environment theme has been the dominating source of (often) conflicting identities in geography for which reason we will briefly draw attention to how determinism, the quantitative revolution and the cultural turn reconfigured the human–environment theme (Turner 2002).

18.3.1 Nature(s) of Determinism and Determining Nature(s)

Debates in the eighteen and early nineteen centuries were largely concerned with environmental determinism (roots from Darwin) versus possibilism and whether culture or nature plays a determining part. Deterministic explanations are often causal and seek to demonstrate how bio-physical factors such as climate, soil and altitude determine social and economic activity. Nature is external, and the domination of nature thesis is particularly inscribed in the enlightenment tradition (Smith 2010). However, in its strictest form the historical-genetic model conceiving nature as the main determinant, never gained full recognition in geography partly because the human–environment theme is poorly understood within isolated and fixed categories that tend to form dichotomies of culture or economy opposed to nature (Christiansen 1967; Fitzsimmons 1989). Environmental determinism suffers from dualist thinking as does much Western philosophy.[4] Such antagonisms have political implications because they involve an attitude of detachment while at the same time holding a perspective of scientific objectivism (Birkeland 1998). Legitimation of geographical knowledge relates to how the object of study is constructed and within the enlightenment tradition, geographers build explanations that objectified nature to be instrumentally used, tamed and exploited (Harvey 1996). In the context of climate change modeling, the environmental determinism thesis can hardly address anthropogenic processes or changes in socio-natural systems, but remain undisputed in the tradition of climate skepticism. Nature dominates culture not vice versa. Yet, the state of much scientific climate change modeling is challenged by the very objectivism of the nature thesis it relies upon, where the concept of nature is often construed in opposition to the concept of culture, either implicitly or explicitly (Lorimer 2012).

18.3.2 Nature(s) and the Quantitative Revolution

Another spatio-temporal tide and wave in the way geographers' deals with human–environment interactions relates to the descriptive tradition/quantitative revolution. During the early and mid-twentieth century positivism became a platform to combat what was regarded speculative science. Universal regularities became a

[4] The concept of nature is often constructed in opposition to the concept of culture, either implicitly or explicitly (Birkeland 1998).

focal point of study (Turner 2002; Hansen and Simonsen 2004). The move from ideographic toward a nomothetic approach in geography reconfigured the human–environment theme, since it could not also encompass environmental determinism. "This obviously implied that the traditional focus of Geography on Human–Environment relationships lost its defining status" (Rasmussen and Arler 2010, p. 38). Ecology was looked upon with much skepticism since it had limited possibilities of quantification, not well suitable for casual and quantitative approaches, despite perhaps the notion of system ecology (Brandt 1999). Subsequently, Kantian geography emphasizing spatial or chorological topographies gained support in favor of the human–environment theme. From the 1960s, however, the focus on environmental problems particularly emerged within the natural sciences, and this in turn gave inspiration to (re)engage with system ecology and human ecology (Rasmussen and Arler 2010). Correspondingly, Zimmerer (2010) discusses nature-society articles from (1911–2010) in the 'Annals of the American Association of Geographers' and shows that the number of articles is nearly as high around the 1960s as during the 1990s and 2000s. The quantitative revolution approaches principles primarily from system ecology; its predictive capacity and analytical certainty (e.g. the assumption of area-biodiversity relations where the size of a geographical area and the successive diversity is a function that equals ecological stability). System ecology, in its original form, suffers from largely being embedded within non-spatial conceptualization, and when spatialized, it remains as implicit constructions of absolute space (Brandt 1999). Quantitative ecological modeling thus reflects mechanistic and general characteristics that presume an equilibrium thesis of which ecological systems tends toward balance and stability (Christiansen 1967; Zimmerer 1994; Hertel et al. 2010).

18.3.3 Nature(s) of Culture and the Linguistic Turn

The cultural or linguistic turn is yet another spatio-temporal tide and wave that influenced human–environment debates, though it never gained much attention in physical geography. According to Birkeland (1998), the rediscovery of the importance of language led to a shift in the relationship between nature and culture favoring socio-spatial formations so that "cultural geography has lost touch with its basic relationship to the concept of nature" (Birkeland 1998, p. 230). According to Fitzsimmons (1989) only a few geographers show interest in the human–environment relationship during the 1980s, and by comparing conceptions of space with conceptions of nature Fitzsimmons demonstrates how geographical thought is imbalanced, not emphasizing the latter. Correspondingly Zimmerer's (2010) analysis of the 'Annals' revealed that articles covering society environmental relations during the 1980s were only half those of the 1960s. Though discursive constructions favored the spatio-chorological tradition the theme never died (Rasmussen and Arler 2010). Interestingly geographical representations of human–environment interfaces decline as a myriad of interdisciplinary fields, from ecological economics, environmental management, to sustainability science bring life to the theme. Discursive constructions, however do share concern for the effects of power for which reason constructionism, particularism and contextual approaches tend to engage in the criticism of the way biophysical environments are construed. Constructions of nature nevertheless face the dilemma of the prison of language: that we can never know if our conceptual construction of nature corresponds to how nature actually appears (Demeritt 2002). Yet, conceptualizations of culture are fundamental to environmental challenges. In addition, material conditions subvert to human emancipation, practices and willpower are changing often in unforeseen ways (Zimmerer 1994; Prigogine 2000; Braun 2006). That claims a strong ontological position about the materiality of nature's construction that implies a rejection of classical divides of subject/object and society/nature dualisms (Latour 1993). Such work has given rise to re-conceptualization of time and space, or concepts like society and nature were privileged ontologies that favored human agency in transforming the environment were challenged. Here relational ontologies comprise 'hybrid' or 'cyborg' forms of human – environment interactions e.g. as co-existing inhabited landscapes, rhizome landscapes or dynamic ecologies, where non-humans agencies are also emphasized (Massey 1999; Whatmore 2006; Lorimer 2012). As we shall see Agent Based Modeling (ABM) may engage in such integrative methodological constructs.

18.4 Spatio-Temporal Figurations and the Geopolitics Modeling

Time-space configurations vary considerably in different sciences. Geologists assemble processes of ecological climatology over millions to billions of years. Evolutionary biologists assemble explanatory power to data stretching thousands to millions of years, whereas many social scientists and economists in particular, are constrained into time-scales of weeks, years and decades due to the practice of discounting (Rasmussen and Arler 2010). These pre-analytic assumptions are fundamental to modeling climate change, and illustrate how the matter of scale and environmental problems are inseparable processes in different time-scales and spatial contexts (Mendelsohn and Dinar 2009). Unifying such a (inter)disciplinary spectrum of different spatio-temporal figurations into representations of climate models poses huge methodological challenges. Moreover, the complexity involved in understanding global climate changes and humans engagement in transforming it, integrate data with causal, functional and intentional explanations (Braun 2006; Rasmussen and Arler 2010). Debate over which data to give explanatory power (agency), are strongly influenced by the time-space figurations adopted. "The way that spatio-temporal processes are studied is strongly influenced by the model of space and time that is adopted" (Raper and Livingstone 1995, p. 262).

18.4.1 Multiple Spatio-Temporalities: Multiple Rationalities

Among Human geographers it is widely acknowledged that space is neither absolute, relative nor relational in itself. Space is produced at one or all scales simultaneously, constituted by the human practices related to it. Some phenomena are represented one dimensional or assumed to be constituted in absolute space as freely unconstrained entities (Harvey 1987). Within human geography it is a general disciplinary assumption that spatio-temporalities are constituted through social processes and interaction with entities with which they mutually constitute entities of indeterminism (Massey 1999). Also in physics and natural sciences such ideas have developed e.g. as biogeochemical ontologies of 'interdependence' (Prigogine 2000; 2008). For authors like Harvey, Thrift or Massey, space and time are integral elements to one another, encompassing multiple spatio-temporalities, constituted by interactions between entities, by which entities are constituted themselves (Harvey 1987; Thrift 1996; Massey 1999). That is, phenomenon e.g. in absolute space cannot be captured with certain representational characteristics of behavior or be given certain actions under which they act rational (humans- or non-humans), without taking into consideration interactions with other spatio-temporal scales. It is essentially another way of saying that linear modeling produce linear results, and such constructs does not capture multiple-spatio temporal interactions. As Massey (1999) points out complexity increases as it becomes apparent that entities conceived epistemologically are also constituted by multiple scales and temporalities inhabited within them (relational space). Thus, we experience conflicts and contradictions between different spatial scales. Subsequently, what seem to be rational in a given scale may not be rational in another; what may be conceived rational in a given spatio-temporal configuration may not in another (Harvey 1987). Human geographers have much to offer conceptual model building in this regard.

By way of example, at one level deforestation is rational to the local farmers in order to expand their production. As biofuel production puts pressure on land use in one location, it may affect e.g. price elasticity elsewhere, not to mention prices on cropland. Thus, relative and redistribute factors are at play. On another scale deforestation is irrational and produces externalities to say tourism, a net-loss of biodiversity (for the 'biotech industries' 'diversity bank'), or climate change mitigation strategies. The problem of land demand is geographically redistributed so in one (relational) scale, afforestation is a rational human action, irrational in another. It therefore becomes more and more evident that contested ideas of 'the market efficiency hypothesis' as equilibrium constructs in climate change or land use modeling is challenged by conflict between different scales, ranging from local to global spatio-temporal figurations "This scale mismatch between an ecosystem (function) and the management set by humans to control or use it constitutes challenges of a theoretical as well as of a more practical nature" (Reenberg 2006, p. 2). This is not to say that we cannot build models that seek to generate scenarios for the futures(s) that fundamentally rely on equilibrium thesis's, but that we may have several equilibrium configurations in different scales, potentially in conflict with one another. It is not the same as different spatio-temporal scales outrage one another and produce a certain kind of status-quo (a new super-equilibrium), with implications of creating new balances or states of stability. This would be like accepting slicing up time and space—ontologically in absolute space. Rather than prioritizing multiple time-scale (in a kind of competition) they are constitutive and contradictory to one another (Massey 1999).

Correspondingly, yet in physics, Prigogine formulate 'a far from equilibrium thesis' assuming that any system is both liner and un-liner and Kleidon (2012) even form a planetary disequilibrium thesis. While Prigogine accepts relations to be causal in some spheres of interaction, he refuses simple linear processes (Prigogine 2004). Causal effects do exist within particular relations in certain spatio-temporal scales. Causal effects exist in multiple versions. But, what is causal in one time and scale may not be causal in another: from small changes that generate large effects (and vice versa), from general processes to contingent events (and vice versa), from local geographical contexts to general or global phenomena (and vice versa). As Cohen and Stevard remarkable note, any system is "Regularities of behavior that somehow seem to transcend their own ingredients" (Cohen and Stevard 1994, p. 232) why concepts of cause relations or equilibrium needs to be viewed dynamically, as always over-floating and interchangeable contingents within and across supposed social and physical spheres. Thus the potential of tipping from domain to domain is always apparent, why emphasis on tipping points, thresholds, abrupt changes or unpredictability, should equally generate deep reflection by the ways in which we assemble conceptual climate modeling (Prigogine 1986; Zimmerer 1994; Kleidon 2012). A world view of such complexity and 'multi-causality' suggest that simple linear and mechanistic scientific approaches sometimes needs to be substituted sometimes supplemented with dialectical reasoning (Harvey 1996).

System thinking refers specifically to the assumption of self-regulating systems, implying that systems possess self- regulating mechanism. Prigogine terms them as dissipative structures, because future is always un-given. Though either Prigogine or Bertalanffy explicated ontological assumptions that established a spatio-temporal theory of human–environmental interactions, they emphasize holism over reductionism and organism over mechanism. If human geography have anything to offer climate change/land use modeling it is to engage in debates on spatial representations that treat concepts like time and space relational, produced by the nature(s) and behavior of entities that inhabit them, rather than time and space themselves are independent of entities they are containing (Massey 1999). To perceive entities as relational constitutions is a fundamentally different approach to spatial modeling of environmental problems, as well as fundamental to the ontological dualism between society-nature (Raper and Livingstone 1995). This has further implications that might be relevant to consider in relation to spatio-temporal tides and waves and pre-analytical assumptions adopted in any 'modeling culture'.

18.5 Modeling Spatio-Temporal Tides and Waves in an Era of the Anthropocene

To define systems, their character and relations to other systems is a journey of geospatial imagination, where we should always question conceptualizations of entities. In this section, the context of space-time relations objectifying nature(s) is examined. Then, spaces of (i)rrationalities are discussed as to different spatio-temporal tides and waves adopted. Lastly, relational ontologies are discussed in the context of Agent Based Modeling. It is argued that Agent Based Modeling engage in such integrative methodological constructs, why we need to develop more appropriate methodological approaches taking into consideration the history of human material interaction.

18.5.1 Anthropogenic Models and Objectifications of Nature(s)

While models focus on the constitution of entities this operate within an object-oriented universe (Brown et al. 2005; Dangermond and Artz 2010). To Massey, approaches in different kinds of representational modeling are "explicitly object-oriented and the objects come before the space-times" (Massey 1999). By way of example, Hertel (2011) concludes that prominent long term agricultural models (e.g. GTAP), tend to treat supply and demand elasticity's based on near term characteristic, why they are not well suited to envisioning long run economic/environmental dynamics. Thus, GTAP tend to adopt short term elasticity characteristics in predicting long term trends. "The tendency to date have been to focus on readily observed, high frequency events, while neglecting some of the important factors which drive the long run dynamics of the system" (Hertel 2011, p. 271). While supply and demand of say corn is aggregated so that a global prize appears as an empirical fact, supply and demand are constituted by multiple heterogeneous characteristics. Though global demand or supply may be aggregated, it is constituted by multiple localized events, responses and capabilities. Interactions are geographically constituted across different spatio-temporalities. Thus relational ontologies accept an aggregated global prize, but are far from reducing it to an objective reality. The surface has its right, but should not dominate at the expense of theory or the philosophy of model building.

The state of much scientific climate change modeling is not only challenged by the objectivism of the social side of climate change, but also the very nature thesis it relies upon. The anthropocene incurs core challenges to the modern (science) understandings of nature as a pure, singular and objective thing separated from human–environmental transformations in multiple scales (subjectivity) (Lorimer 2012). We can hardly talk about anthropogenic climate change adaption or mitigation (Oreskes 2004; Steffen et al. 2011; Cook et al. 2013) and at the same time argue for a purely objective nature (of science) opposed to culture. Thus, the anthropocene challenges the modern science-politics settlement, where natural science speaks for an objective nature (Latour 1993; Demeritt 2002; Lorimer 2012). The material/relational human–environment ontologies force us to develop a move from a mechanical view of nature as external towards more dynamic conceptualizations of human-nature interfaces in climate (land use) model building. Mutual construction implies a rejection of classical divides of subject/object and society/nature dualisms central to anthropocentrism and essentialist assumptions of conceptual models (Birkeland 1998). Yet, subject-object and society-nature reunioning have to be conceptualized in much climate change modeling and suggest that we engage in explaining entities of reductionism, indeterminacy, path-dependency or irreversible processes that our conceptualizations arrive from. The following is to argue that debate, questions, rivalry and 'tentative' struggles over problems of 'rhizome interfaces' that both natural and social sciences have in common, provide a simulative platform to engage in the challenges of reimagining the multiple dynamics shaping conceptual climate change modeling (Whatmore 2006; Massey 2006).

18.5.2 Spaces of (I)rationalities and the Equilibrium Thesis: Mimicking the Quantitative Revolution

The systems of geospatial imagination are often organized hierarchally with related systems and subsystems, and accompanied interactions weather causal, linear, abrupt or unpredictable. Systems and models are closely related and widely used as representations of reality in natural and parts of the social sciences. Yet, the terminology of models is extremely diffuse and preoccupied with much skepticism in social sciences, perhaps except for economics (Rasmussen and Arler 2010). Much conceptual climate model building reduces human-nature interactions to questions of economic calculation. Though economics is important, a lot more than economics is going on in human–environmental interfaces, and economy has a limited explanatory power in itself (Urry 2011). By way of illustration mainstream economics have historically treated energy as a free good. In fact any natural resource has been considered a free input to economic growth. Thus energy or material resources are first treated as a free good (despite we live on a finite planet), then like a commodity as any other, not perceiving the dynamic relations and the material constrains they rely upon (Kock 2012). Discussions on environmental determinism precisely engage in portraying fixed entities (like resources) and weighting them as cultural and natural factors (spatial and temporal figurations) in competition to one another, and thereby a sort of hierarchy also weighting disciplinary knowledge like economic factors opposed to cultural knowledge, implicitly or explicitly (Demeritt 2002; Lorimer 2012). The problem of cause is how fixity, stability and equilibriums explain change and socio-ecological transformations. Models are not able to deal with all uncertainties in complex non-linear dynamics systems, nor interconnections between systems and subsystems, hence we should comprehensively and critically question how models are assembled and what type of knowledge for what purpose that arises from them (Zimmerer 1994). When Monica Turner (2005) for example advocates that landscape ecology should "develop a more mechanistic understanding of the relationship between pattern and process" (Turner 2005, p. 319) it contrasts Zimmerer's (1994) advocacy for landscape ecology and its effort in understanding biophysical environments also under non-equilibrium conditions. Geographers have long challenged the equilibrium and stability thesis. Thus human geographers are in a good position to critically scrutinize e.g. the sub politics of elasticity parameters construed as well as consequences of spatio-temporal figurations associated with it (Hertel et al. 2010).

Commitment to a theory of knowledge, according to which any phenomenon natural or social is to be explained through systems of laws and causalities mimicking the quantitative revolution, does not fit well with the social dimension of climate change, irreversible processes nor abrupt changes (Rasmussen and Arler 2010; Kleidon 2012). According to this perspective climate change can be instrumentally adjusted as a form of global technocratic climate management (Urry 2011). In this sense, the gradualist perspective of climate models carries references to the quantitative revolution. In recent years the human–environment theme dominated by gradualist approaches to climate change seem to convent a new form of positivism' in much climate change modelling. If it is assumed that a dialectic approach comprehends the complexity of socio-spatial and economic-ecological processes, this, in turn, will make us recognize that environmental/social problems mutually interact, are spatially distributed, and produce different effects in different spatial scales (Harvey 1996). Interdisciplinary approaches seem fundamental to the analysis of wicked problems, multi-complex and multivariable interactions associated with climate change, and the methodological challenges associated with rhizome ontologies, giving different kinds of data agency in models.

18.5.3 Agent Based Modeling and Rhizome Ontologies

For various reasons Agent Based Modeling (ABLUM) has received much attention in recent years. First of all ABM offers a methodological approach integrating human decisions e.g. on land use, based on monetary and non-monetary 'calculations' from particular agents as a starting point. Particularly the integrative approach to modelling individual decision making, interactions, and social non-monetary processes that dynamically link to environmental processes have been considered a central advantage (Brown et al. 2005; Turner 2005; Matthews et al. 2007; Barton et al. 2010). Privileged ontologies that favor human agency in transforming the environment has long been challenged by much human geographical work that also gives non-humans agency (Whatmore 2006; Lorimer 2012). To give non-humans agency is precisely what ABM does (Turner 2005). These actor-networks represented, originate from the field of artificial intelligence and individual based modeling. Accordingly actors are given agency that simulates certain characteristics so they interact both with each other and their environment. Thus, more than human interactions (Whatmore 2006), in AGM is modeled in ways that agents can take and change decisions based on interactions with other agents and the environment. As responses dynamically changes, Whatmore's notion of more than human agency become integrated into the framework of ABM, and implies that the environment is an agent too.[5] Whether human or non-human they become subject to subjectivity. This makes a whole lot of difference to modeling dynamics, yet it seems to promulgate new forms of positivism in modeling of land use/climate change. "The behavior of the whole system depends on the aggregated individual behavior of each agent. This allows the influence of human decision-making on the environment to be incorporated in a mechanistic and spatial explicit way, also taking into account social interaction, adaption and decision-making at different levels" (Matthews et al. 2007, p. 1448). Consequently, ABM operates in an objective, oriented universe, with ever more sophisticated aggregation matrixes of interactive events, not encompassing that the whole is more than the sum of its parts. Non-humans are multiple automated agents, yet regarded autonomous in the software language, created with rules for formulating decisions while interacting with the environment: "They are instantiated ('activated') on virtual landscapes and allowed to act and interact over time without intervention by the researcher" (Barton et al. 2010, p. 5383). AGM gives agency as a form of new positivism, where 'deterministic reductionism' is assembled to simple models as a sum of perceived components (Urry 2011). As to Whatmore's (2006) configurations of 'more than human' AGM opens the journey for an 'objective structuralism' as mechanistic arrangements of these human and non-human configurations, while ignoring the very subjectivity model culture they are design from. ABM however, is in a position to integrate different rationalities in play in different scales, there relational dynamics, contradictions and subordinate characteristics. If greater reliability of models, they should be able to run forwards as well as backwards simultaneously; should be able to start at different points in time, space and scale, and derive the same results. Complexity increases, however, as it is argued that models and their seemingly neutrality (O'Sullivan 2005; Barton et al. 2010) is challenged by the very idea that models are agents themselves.

18.6 The Geopolitics Models (Continued)

> Essentially, all models are wrong, but some are useful. . . . [T]he practical question is how wrong do they have to be to not be useful (Box and Draper 1987, p. 74).

[5] Modeling elephants' responses to climate change or urbanization.

In this final section we discuss some of the contributions the Human side of geography can to offer climate change modeling. The terminology of models and modeling is extremely diffuse and human geographers are in a good position to make a leading contribution as to the spatio-temporal implications associated with it. The famous quotation by Box and Draper (1987) spurred a vivid debate as to the use (fullness) of models. Today it is more relevant than ever as the emerging state of much modeling integrate the social side of (inter) action. The ways we perceive the world hugely influence how we act (and vice versa). By analogy Clegg and Hardy state that the normative connections embed “ways of seeing which act back on and reflect existing ways of seeing” (Clegg and Hardy in Alvesson and Sköldberg 2009, p. 248). To frame models as objective unbiased observation of human–environment interaction is to ignore the power relations inherent in any research agenda. Power relations form the very interpretative categories the models are designed from (Demeritt 2002). The (geo)politics of climate change models have bearings to actions taken and therefore ‘interact’ with the environment itself. Models are then an agent in it-self that act and interact with other agents, and so take part in shaping new meteorological and socio-ecological futures. Models then also become a political tool that helps construe different scenarios to take decisions upon. Thus, modeling different climate future(s) or land use scenarios exactly is value laden representations with an intention to impact other agents. In that sense climate models convert into sophisticated forms of geopolitics and geo-engineering. The gradual perspective in much model building assume that better technical management of human–environment relationship, e.g. through better and more accurate modeling, is needed and enhance the knowledge decision rely upon (Urry 2011). Thus models are also emergences of geo-engineering or planetary management, that commit model scenarios of adjustment, themselves taking part in modifying metrological future(s). Political settlements of modeling future(s) where scientists speak of an objective nature, providing facts about that objective nature, and politicians ask for facts to take decisions upon, makes Haraway conceptualize dynamics of charm as a sense of ‘response-ability’, by which different kinds of agents have adaptive transformative and resistive capabilities, that affect and is affected by others actions in that relationship. Such hybrid ontologies do not only pose questions to models as neutral representations of reality and illustrate the need for critical approaches to the intentional content of model results, it also unfold that it is politically insufficient to analytically constrain debates into the endless dialectics, hybridity and uncertainty involved in future climate scenario building (Lorimer 2012). The social side of land use/climate change models is cruical for designing future scenarios and associated desicions based on such models, why another approach to modeling is required as to the politics of modeling, not least an awareness about the limits to what policy input these models arrives from and the politics produced. We therefore also need to ask what kind of climate or land use modeling for what kind of socio-ecological future? Who decide the culture of modeling construction, on what grounds and through what processes, and how do they influence decision making processes? Agent Based Modeling has brought much dynamism into the modeling culture. It will be interesting to observe whether or not (agent based) models will also be able to take into consideration themselves as agents and the geopolitical implications hereof.

Acknowledgments I would like to thank Henrik Toft Jensen, Associate Professor at Roskilde University, Department of Environmental, Social and Spatial Change, and Mette Termansen, Professor, Aarhus University, Department of Environmental Science, for insightful contributions and lively discussion, as well as Andrew Crabtree for proof reading.

References

Alvesson M, Sköldberg K (2009) Reflexive methodology: new vistas for qualitative research. Sage, London

Barton M, Ullah I, Bergin S (2010) Land use, water and Mediterranean landscapes: modelling long-term dynamics of complex socio-ecological systems. Phil Trans R Soc A 368:5275–5297. doi:10.1098/rsta.2010.0193

Birkeland IJ (1998) Nature and the 'cultural turn' in human Geography. Norwegian J Geogr 52:221–228. doi:10.1080.0029195980855.4.2

Box GE, Draper NR (1987) Empirical model building and response surfaces. Wiley, New York

Brandt J (1999) Geography as landscape ecology. Danish J Geogr 1:21–33, http://img.kb.dk/tidsskriftdk/pdf/gto/gto_si01-PDF/gto_si01_72554.pdf

Brown D, Riolo R, Robinson D, North M, Rand W (2005) Spatial process and data models: toward integration of agent-based models and GIS. J Geogr Syst 7:25–47. doi:10.1007/s10109-005-0148-5

Braun B (2006) Towards a new earth and a new humanity: nature, ontology, politics. In: Castree N, Gregory D (eds) David Harvey: a critical reader. Blackwell, Oxford, pp 191–222

Christiansen S (1967). Geografi som "human økologi". Danish J Geogr Bind 66:24–35

Cohen J, Stevard I (1994) The collapse of Chaos: discovering simplicity in a complex world. Penguin, New York

Cook J, Nuccitelli D, Green SA, Richardson M, Winkler B, Painting R, Skuce A (2013) Quantifying the consensus on anthropogenic global warming in the scientific literature. Environ Res Lett 8:1–7. doi:10.1088/1748-9326/8/2/024024

Crutzen PJ (2002) Geology of mankind. Nature 415:23. doi:10.1038/415023a

Dangermond J, Artz M (2010) Climate change is a geographical problem. ESRI, Redlands

Demeritt D (2002) What is the 'social construction of nature'? A typology and sympathetic critique. Prog Hum Geogr 26:767–790. doi:10.1191/0309132502ph402oa

Elling B (2003) Rationality and the environment, decision-making in environment and assessment. Earthscan, London

Fitzsimmons M (1989) The matter of nature. Antipode 21(2):106–120. doi:10.1111/j.1467-8330.1989.tb00183.x

Griggs D, Stafford-Smith M, Gaffney O, Rockström J, Öhman M, Shyamsundar P, Noble I et al (2013) Sustainable development goals for people and planet. Nature 495:306–307. doi:10.1038/495305a

Hansen F, Simonsen K (2004) Geografiens Videnskabsteori – en introducerende diskussion. Roskilde University Press, Roskilde

Harvey D (1969) Explanation in geography. Edward Arnold, London

Harvey D (1987) The condition of postmodernity. Blackwell, Oxford

Harvey D (1996) Justice, nature and the geography of difference. Blackwell, Oxford

Hertel WT (2011) The global supply and demand for agricultural land in 2050: a perfect storm in the making? Am J Agric Econ 93(2):259–275. doi:10.1093/ajae/aaq189

Hertel WT, Rose S, Tol RS (2010) Land use in computable general equilibrium models: an overview. In: Hertel WT, Rose S, Tol RS (eds) Economic analysis of land use in global climate change policy. Routledge, Abingdon

IPCC (2013) Fifth assessment report climate change 2013. The physical science basis summary for policymakers, IPCC

Kleidon A (2012) How does the Earth system generate and maintain thermodynamic disequilibrium and what does it imply for the future of the planet? Phil Trans R Soc A 370:1012–1040. doi:10.1098/rsta.2011.0316

Kock M (2012) Capitalism and climate change, theoretical discussion, historical development and policy responses. Palgrave MacMillan, London

Latour B (1993) We have never been modern. Harvard University Press, Cambridge

Lorimer J (2012) Multinatural geographies for the Anthropocene. Prog Hum Geogr 36(5):593–612. doi:10.1177.030913251143.3.2

Massey D (1999) Space-time, 'science' and the relationship between physical geography and human geography. Roy Geogr Soc 24:261–276. doi:10.1111/j.0020-2754.1999.00261

Massey D (2006) Landscape as provocation. Reflections on moving mountains. J Mater Cult 11:33–48. doi:10.1177.135918350606.9.1

Matthews RB, Gilbert GN, Roach A, Pohill GJ, Gotts NM (2007) Agent-based land-use models: a review of applications. Landsc Ecol 22:1447–1459. doi:10.1007/s10980-007-9135-1

McNeill JR (2000) Something new under the sun. WH Norton and Company, New York

Mendelsohn R, Dinar A (2009) Land use and climate change interactions. Annu Rev Resource Econ 1:309–332. doi:10.1146/annurev.resource.050708.144246

O'Sullivan D (2005) Geographical information science: time changes everything. Prog Hum Geogr 29(6):749–756. doi:10.1191/0309132505ph581pr

Oreskes N (2004) Beyond the ivory tower - The scientific consensus on climate change. Science 306:1686. doi:10.1126/science.1103618

Prigogine I (1986) Science, civilization and democracy: values, systems, structures and affinities. Futures 18(4):493–507. doi:10.1016/0016-3287(86)90074-1

Prigogine I (2000) The future is not given in society or nature. New Perspect Q 17(2):35–37. doi:10.1111.0893.7850.00262

Prigogine I (2004) Beyond being and becomming. New Perspect Q 21(4):5–12. doi:10.1111/j.1540-5842.2004.00689.X

Prigogine I (2008) Is the future given? changes in our description of nature. In: Sidharth BG (ed) A century of ideas. Springer, Heidelberg, pp 65–75

Raper J, Livingstone D (1995) Development of a geomorphological spatial model using object-oriented design. Int J Geogr Inform Syst 9:359–383

Rasmussen K, Arler F (2010) Interdisciplinarity at the human-environment interface. Danish J Geogr 110(1):37–45. doi:10.1080.00167223.2010.10669495

Reenberg A (2006) Land systems research in Denmark: background and perspectives. Danish J Geogr 106(2):1–6. doi:10.1080.00167223.2006.10649552

Smith N (2010) Uneven development, nature, capital and the production of space, 3rd edn. Verso, London

Steffen W, Grinewald J, Crutzen P, McNiel J (2011) The Anhtropocene: conceptual and historical perspectives. Phil Trans R Soc A 369:842–867. doi:10.1098/rsta.2010.0327

Thrift N (1996) Spatial formations. Sage, London

Turner B (2002) Contested identities: Human-environment geography and disciplinary implications in a restructuring academy. Ann Assoc Am Geogr 92(1):52–74. doi:10.1111.1467.8306.00279

Turner M (2005) Landscape ecology: what is the state of the science? Annu Rev Ecol Evol Syst 36:319–344. doi:10.1146/annurev.ecolsys.36.102003.152614

Urry J (2011) Climate change and society. Polity, Cambridge

Whatmore S (2006) Materialist returns: practicing cultural geography in and for a more-than-human world. Cult Geogr 13:600–609. doi:10.1191/1474474006cgj377oa
Zimmerer KS (1994) Human geography and the "new ecology": the prospect and promise of integration. Ann Assoc Am Geogr 84(1):108–125. doi:10.1111/j.1467-8306.1994.tb01731.x
Zimmerer KS (2006) Humboldt's nodes and modes of interdisciplinary environmental science in the Andean world. Geogr Rev 96(3):335–360. doi:10.1111/j.1931-0846.2006.tb00255.x
Zimmerer KS (2010) Retrospective on Nature–Society geography: tracing trajectories (1911–2010) and reflecting on translations. Ann Assoc Am Geogr 100(5):1076–1094. doi:10.1080.00045608.2010.523343

Chapter 19
Modelling of Dynamic Relationship Between Socio-Economic Changes and Deforestation in Kolasib District of Mizoram

Rintluanga Pachuau and H. Lalchamreia

Abstract Kolasib district is located in the northern most part of the state of Mizoram. The state of Mizoram is lying in the far north east corner of India. In order to analyze the dynamic relationship between socio-economic changes and deforestation in Kolasib district of Mizoram, regression model has been adopted. The main purpose of applying regression model is to establish a functional relationship between the two variables (socio-economic and forest and non forest land use), which will enables us to determine the magnitude of one variable as determined by the others. The estimation is made on the basis of the cross sectional data. An attempt is made to estimate the log-linear regression change model; it is observed that the estimated coefficient is significantly negative indicating that change in the socio-economic is negatively responded to by forest cover. So, it may be concluded that the Logarithmic Model being proposed here to fit the forest land use change is significantly valid in case of time series as well as cross section data; and hence it may be considered as appropriate model to fit the observed forest land use change.

Keywords Accessibility • Deforestation • Land use • Model • Socio-economic changes

19.1 Introduction

Model has been defined as an idealized and structured representation of the real world (Hussain 2007). According to Ackoff et al. (1962), "a model may be regarded as a formal presentation of a theory or law using the tools of logic, set theory and mathematics". The function of model is to illustrate the theory and summarize the

R. Pachuau • H. Lalchamreia (✉)
Department of Geography and Resource Management, School of Earth Sciences, Mizoram University, Aizawl, Mizoram, India
e-mail: d.lcr@rediffmail.com

M. Singh et al. (eds.), *Climate Change and Biodiversity: Proceedings of IGU Rohtak Conference, Vol. 1*, Advances in Geographical and Environmental Sciences,
DOI 10.1007/978-4-431-54838-6_19,

complex relations of the real world in a more precise way. When well developed theory is represented by certain mathematical equations, this set of equations in turn represents some real world system; such representation is known as mathematical model (Singh and Singh 1991).

Since the beginning of human civilization, deforestation already took place. Humans have controlled fire and domesticated plants and animals; they have cleared forests to wring higher value from the land (Lambin et al. 2003). About half of the ice-free land surface has been converted or substantially modified by human activities over the last 10,000 years. A recent study estimated that undisturbed (or wilderness) areas represent 46 % of the earth's land surface (Mittermeier et al. 2003). Forests covered about 50 % of the earth's land area 8,000 years ago, as opposed to 30 % today (Ball 2001). Agriculture has expanded into forests, savannas, and steppes in all parts of the world to meet the demand for food and fiber. Agricultural expansion has shifted between regions over time; this followed the general development of civilizations, economies, and increasing populations (FAO 2001).

Deforestation is primarily confined to developing countries, and mainly in the tropics (Myers 1994). There is growing concern over shrinking areas of tropical forests (Barraclough and Ghimire 2000). During 1960–1990, the loss of natural forest area in the tropics was estimated at 450 million ha; of which the loss during 1981–1990 was 154 million ha (FAO 1993; Singh and Jans 1995). Contrary to the trend in tropical countries, the deforestation rate per unit population in India is one of the lowest among all the major tropical countries. During 1990–2000, the annual rate of loss aggregated for the entire country was estimated to be 0.1 % (FSI 1999). However, regional differences were significant: the NE Region lost 2,000 km^2 of forest area in spite of the Forest Conservation Act, 1980; the extent of deforestation and forest degradation is also very high.

Mizoram is one of the seven states in the North East Region of India and it presents a miniature image of North Eastern Region in terms of forests and socio-economic diversity. As per FSI Report, out of the total land area of 2.1 million ha, forests occupy 1.9 million ha (or 88.6 %); the forest loss during 1990–2005 was 87,000 ha; 57 % of its forests had crown density less than 40 % and 0.3 million ha (19 %) were subject to annual shifting cultivation (FSI 2005). However, deforestation in Mizoram is human induced and happened over time and space.

19.2 Study Area

Kolasib district is situated in the northern most part of Mizoram between 24°31′ 14.43″ and 23°51′ 15.13″N latitudes and 92°31′ 46.92″ and 92°54′ 11.40″E longitudes. The district falls within the Survey of India Topo sheet Nos. 83D/11, 83D/12, 83D/14, 83D/15, 83D/16, 84A/9 and 84A/13 (see Fig. 19.1). It is bounded by Cachar district of Assam on the north, Aizawl District of Mizoram on the south

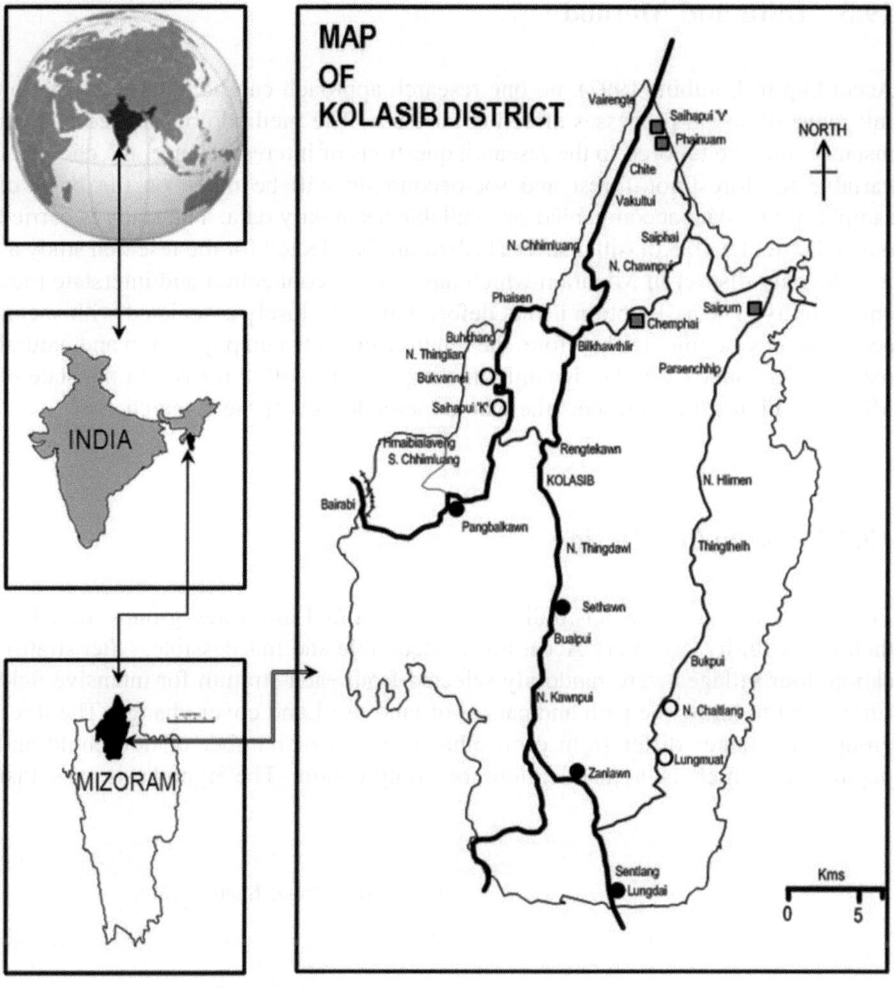

Fig. 19.1 Location map

and east, on the west by Karimganj district of Assam and Mamit district of Mizoram. The total aerial extent of the district is 1,382 km^2. There are 83,054 population as per 2011 census, having a density of 60 persons/km^2 against the density of the Mizoram state—52 persons/km^2. According to 2011 census, the rural population constitutes 43.78 %, and urban population comprises 56.22 % of the district population. The tribal population constitutes 89.78 % of the district population. The literacy rate is 91.49 %. There are 48 villages in the district as per 2011 census. Majority of the population depends on agriculture and other land base occupation.

19.3 Data and Method

According to Lambin (1997), no one research approach can hope to elucidate the full range of social processes affecting land use. The methodologies used in each instance must be tailored to the research questions of interest. Hence, the choice of variable for forest/non-forest and socioeconomic will be made on the basis of sample survey data accompanied by available secondary data. The study is carried out at district level. Kolasib district of Mizoram is selected for the research study as it is the only district in Mizoram which has railway connection and interstate road connectivity. The assumption is that deforestation is closely associated with socio-economic development. Therefore, the relationship between population and natural resources is supposed to be dynamic in comparison to other district in the state of Mizoram. Figure 19.2 presents the whole methodology of the research.

19.3.1 Sampling Design

The census villages of Kolasib district were stratified into three groups, based on their accessibility. Viz Very Accessible, Accessible and Inaccessible. After stratification, four villages were randomly selected from each stratum for intensive field survey and to know the path and causes of land use Land cover change. The three groups of villages differ from each other in terms of number of household and population, which is a justification for stratification. The hypothesis was that

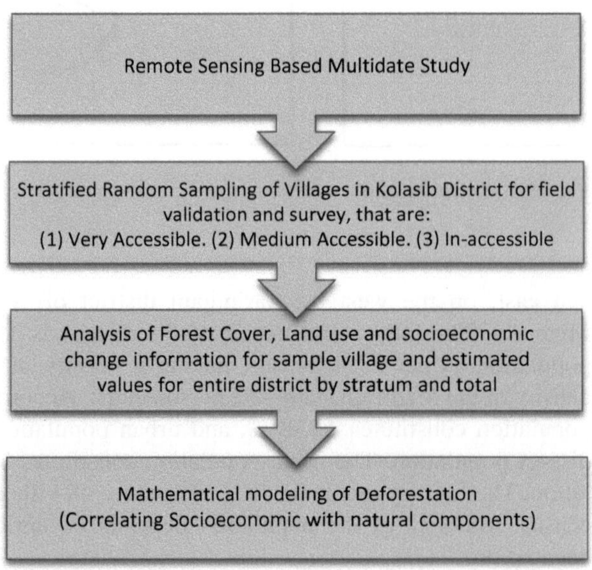

Fig. 19.2 Methodology chart

deforestation closely relates economic development, which tends to follow accessibility classes. The survey design is Stratified Random Sampling (Cochran 1963). Based on accessibility criteria, three strata have been distinguished. The areal extents of Forest and non-forest cover were quantified with the help of community participation by using IRS-IC/ID LISS-III 2002 for twelve sampled villages respectively.

19.3.2 Data Source

The primary data pertaining to socio-economic variables like population, household, livelihood and land use pattern are collected from 12 sampled villages for two dates by using specific schedules and structured questionnaire. These data were collected for two date's time i.e. 1997 and 2007 for in depth analysis of changes in forests and socio-economic condition at village level, which determine the total assumption of the whole district. The secondary data pertaining to various population parameters is collected from Census of India (1981, 1991, 2001), various state and central government of India reports. Forests related information including village and district boundaries is also collected from Survey of India topo-sheets and Forest Survey of India reports. The areal extents of forest and non-forest areas are obtained by using IRS-IC/ID LISS-III 2002 procured from National Remote Sensing Agency (NRSA), Hyderabad, India.

19.4 Model Specification

In order to analyze the relationship between socio-economic changes and deforestation, regression model has been adopted. The main purpose of applying regression model is to establish a functional relationship between the two variables, which will enable us to determine the magnitude of one variable as determined by the others. The following functional relationship is proposed to represent the interrelationships between the socio-economic changes and deforestation in the study area, which is Kolasib district of Mizoram.

$$Y = f(\mathrm{x}) \tag{19.1}$$

where 'Y' is the ratio of forest land use to total land area, and 'x' represents the explanatory variable (predictor) of forest land use. Least squares method is applied to estimate this functional relationship and the calculated R-square (also called coefficient of determination) which measures the ratio of the dependent variables (Y) explained by the explanatory variables or predictor (X). The ratios of forest land use, as represented by Y in Eq. (19.1), are given by the following expression:

(i) $Ratio\ of\ Forest\ Cover\ (F) = \frac{Total\ Forest\ Area}{Total\ Land\ Area}$

(ii) $Ratio\ of\ Dense\ Forest\ (DF) = \frac{Total\ Area\ of\ Dense\ Forest}{Total\ Forest\ Area}$

(iii) $Ratio\ of\ Moderate\ Forest\ (MF) = \frac{Total\ Area\ of\ Moderate\ Forest}{Total\ Forest\ Area}$

(iv) $Ratio\ of\ Open\ Mix\ Forest\ (OP) = \frac{Total\ Area\ cover\ by\ Open\ Mix\ Forest}{Total\ Forest\ Area}$

Similarly, the explanatory variable, as represented by X in Eq. (19.1), comprises of population (*P*), number of Households (*HH*) and Livelihood Change Index (*LCI*). The present analysis considered the impact of the changes population, in line with the study by Singh et al. (2012), and *LCI*. Livelihood change index is an index specially constructed to represent the changes in the livelihood of the households in the study area. The main objective of constructing this index is to emphasis the relevance of adopting livelihood which is considered as having lesser pressure on forest area. It is represented by the following expression:

$$LCI = \frac{M + Pr + FrM + L + JB}{5} \tag{19.2}$$

where

$M = 1 - \frac{Number\ of\ Jhumia}{Total\ Households}$ is the shifting indicator of households from the practice of jhum cultivation and $FrM = 1 - \frac{No.\ of\ Families\ involve\ in\ Marketing\ of\ Forest\ Product}{Total\ Households}$ is the shifting indicator of families from forest business to other; while *Pr, L* and *JB* are the ratio of permanent cultivator, Livestock and Job & business respectively. The *LCI* index will increase with the decrease in the number families who depend on forest and its produce for livelihood. That is, increase in LCI will reflect the decrease in further dependency of the population on natural forest.

The next problem is to choose suitable regression model to fit the observed data. Theoretically, the model which shows better goodness of fit as represented by R-square should be chosen for making conclusion. After simulation of various regression models, the performance of log-linear regression model with zero intercept was found to be most satisfactory to fit our field data. It is given by the following equation:

$$Y = X^{\beta} u \tag{19.3}$$

where u is the random error term. The model presented in Eq. (19.3) can be transformed into linear model by simple logarithmic transformation (so it is called log linear) as follows:

$$\log Y = \beta \log X + \log \mu.$$

Or,

$$\log Y = \beta \log X + \mathrm{e} \tag{19.4}$$

where e = logμ is the error term. The main advantage of this model over any other functional form is its assumption of constant elasticity (i.e. β). Elasticity is defined as the degree of responsiveness of the dependent variable to the changes in the independent variable (or predictor). Symbolically, it is given by

$$Elasticy\ (\beta) = \frac{Proportionate\ Change\ in\ Y}{Proportionate\ Change\ in\ X}$$

If $\beta > 1$, one percent increase (decrease) in *X* will result in an increase (decrease) in *Y* by more than proportionately; if $\beta < 1$, one percent increase (decrease) will result in an increase in *Y* by less than proportionately; while the reverse will be true if the elasticity is negative.

19.5 Results and Discussion

The log-linear model, given in Eq. (19.4), of the forest land use on each predictor or explanatory variable is estimated using each date of 1997 and 2007, for which data have collected, and overall period separately. But the result shows more or less similar pattern, it is decided to adopt the result of the combined samples (1997 and 2007). As the estimation is made on the basis of the cross sectional data, there is no theoretical problem in combining the data of 1997 and 2007; it is rather suggested, on a priori ground, because of the gains in degrees of freedom.

Table 19.1 presents the result of the regression equation of various forest land uses on population and shows all coefficients are significant at all level, which is associated by satisfactory goodness of fit that is R-square, which ranges from 0.74 in case of Open/Mix forest and to 0.87 in case of dense forest. It would be observed from this table that population growth is negatively responded to by the percentage of the various forest land use classification. Open/mix forest shows the lowest coefficient at −0.116, and will be the least affected by population growth; while dense forest will be the most affected by population growth as it has shown the largest coefficient (absolute term). The general conclusion that can be drawn from Table 19.1 is that population growth shall significantly results in deforestation.

Table 19.1 Estimated regression: Model-I

Predictor: log (P)

Dependent variables	Coefficient	Std. error	t-ratio	Sig.	R-Square	Adj. R-Square
log (F)	−0.131***	0.015	−8.58	0.000	0.76	0.75
log (DF)	−0.255***	0.020	−12.54	0.000	0.87	0.87
log (MF)	−0.223***	0.024	−9.46	0.000	0.8	0.79
log (OP)	−0.116***	0.014	−8.06	0.000	0.74	0.73

***, ** & * indicates significant at 1 %, 5 % & 10 % respectively

Table 19.2 Estimated regression Model – II

Predictor: log (HH)						
Dependent variables	Coefficient	Std. error	t-ratio	Sig.	R-Square	Adj. R-Square
log (F)	−0.173	0.020	−8.72	0.000	0.77	0.76
log (DF)	−0.335	0.028	−12.06	0.000	0.86	0.86
log (MF)	−0.291	0.032	−8.97	0.000	0.78	0.77
log (OP)	−0.153	0.019	−8.12	0.000	0.74	0.73

***, ** & * indicates significant at 1 %, 5 % & 10 % respectively

Table 19.3 Estimated regression Model – III

Predictor: log (LCI)						
Dependent variables	Coefficient	Std. error	t-ratio	Sig.	R-Square	Adj. R-Square
log (F)	0.506***	0.158	3.214	0.007	0.46	0.42
log (DF)	0.769***	0.238	3.237	0.001	0.47	0.42
log (MF)	0.845***	0.199	4.255	0.001	0.60	0.57
log (OP)	0.305**	0.117	2.601	0.023	0.36	0.31

***, ** & * indicates significant at 1 %, 5 % & 10 % respectively

Table 19.2 presents the effect of the increase in the number of households on the forest land use change. As the estimated coefficient are significant at all levels, with satisfactory R-square, and are all negative, the changes in the number of households (increase or decrease) will result in the forest land use change in the reverse direction. It should be noted that the increases in either of the two variables of population or the number of household represents population growth. Therefore, the result of the two estimated regression models, presented in Tables 19.1 and 19.2, led to the conclusion that population growth will, invariably, have negative impact on natural forest (i.e. deforestation).

Table 19.3 shows the different picture from the previous two models that all estimated coefficient are significant and positive, though it shows relatively lower value of R-square. Forest land use in the study area, as a percentage of the total area, will positively responded to the livelihood changes. This result suggests that the shifting of the population from forest based livelihood (like jhumming, marketing of forest products, etc.) to forest-friendly activities (like permanent farming, jobs & business, etc.) will have unequivocal impact on the preservation of forest and afforestation

Table 19.4 presents the elasticity of forest land use on the changes in the population, number of households and livelihood (simply the coefficients of the logarithm of these variables). It would be observed from this table that all elasticities of forest land use on population increase turned out to be negative and less than unity, the result shows population increase will be associated with the decrease in total forest coverage, but less than proportionately. In other words, one per cent increase in population will result in less than one per cent decrease in forest cover. As it shows the lowest elasticity at −0.116, open mix forest will be the least

Table 19.4 Estimated elasticity of forest land use to socio-economic changes

Variable/indicators	Forest cover	Dense forest	Moderate forest	Open/mix forest
Population	−0.131***	−0.255	−0.223	−0.116
Number of households	−0.173***	−0.335	−0.291	−0.153
Livelihood change index	0.506***	0.769	0.845	0.305

***, ** & * indicates significant at 1 %, 5 % & 10 % respectively

affected by the population increase; while dense forest, which has elasticity of −0.255, will be the most affected by population increase. The same trend is being observed in case of the elasticities of forest on households increase or decreases, with slightly higher magnitudes. It would be, therefore, concluded that forest is more volatile to changes in the number of households than population increase. We can make inference here that population increase and its associated fragmentation of families, due to small family norm encouraged by the government, will result in demand of more area for cultivation and hence aggravating deforestation.

At the same time, all estimated elasticities with respect to the indicator of livelihood changes turned out to be positive and significant. As it was given in the preceding section, the increase in the value of LCI represents the development of the population by adopting better and more sustainable livelihood, abolishing traditional practice of jhumming, by following settled farming, job & business, etc. That is to say that eco-friendly development in the study area will have positive impact on the natural forest.

19.6 Time Series Data Validity of the Model

Another academic interest is to examine the validity of cross-sectional model in case of time series data. An attempt is made in this section to analyze the performance of log-linear change model, which successfully fit our cross-sectional village data, in time series. The time series data of total forest cover in Kolasib district as provided by Forest Survey of India between 2001 and 2011 along with Population Data provided by Directorate of Census Operation Mizoram are presented in Table 19.5. As the Directorate of Census Operation does not provide Year-Wise population figure, estimation is made using the decadal growth rate calculated by Census Department, Mizoram.

Using the information on forest cover and population in Table 19.5, an attempt is made to estimate the log-linear regression model, as given in Eq. (19.4), and the estimated regression equation is presented in Table 19.6. It would be observed from this table that the estimated coefficient is significantly negative indicating that change in the population is negatively responded to by forest cover. This may be taken as an enhancement of the previous analysis of the cross-section village data. So, it may be concluded here that the Logarithmic Model being proposed here to fit the forest land use change is significantly valid in case of time series as well as cross section data; and hence it may be considered as appropriate model to fit the observed forest land use change.

Table 19.5 Population and forest cover in Kolasib district from 2001 to 2011

	Population			Forest cover	
Year	Rural	Urban	Total	Area (km^2)	Area (%)
2001	29,461	36,499	65,960	1,322	95.66
2002	30,151	37,519	67,669	1,333	96.45
2003	30,857	38,567	69,424	1,344	97.25
2004	31,579	39,645	71,223	1,305	94.43
2005	32,318	40,752	73,070	1,266	91.61
2006	33,075	41,891	74,966	1,274.5	92.22
2007	33,849	43,061	76,910	1,283	92.84
2008	34,641	44,264	78,906	1,291.5	93.45
2009	35,452	45,501	80,954	1,300	94.07
2010	36,282	46,773	83,055	1,255.5	90.85
2011	37,132	48,079	85,211	1,211	87.63

Note: Year-wise population is generated using the estimated decadal growth rate of population Census between 2001 and 2011 (i.e. Rural: 23.41 % & Urban: 27.94 %)
Source: (i) Directorate of Census Operation, Mizoram, Aizawl. & (ii) FSI (2001, 2003, 2005, 2009, 2011) & Sample survey-2007 for forest cover

Table 19.6 Estimated regression equation of forest area (%) on population

Coefficient	Std. error	t-ratio	Sig.	R-Square	Adj. R-Square
−0.00622	0.001	−7.910	0.000	0.86	0.85

Equation: log(forest) = b log(population)

Table 19.7 Forecasted forest cover (%) in Kolasib district using the estimated regression model of log (forest) = −0.00622 log(population)

Year	Population	Actual forest area (%)	Forecasted forest area (%)	Error (%)
2001	65,960	95.66	93.33	2.33
2002	67,669	96.45	93.32	3.14
2003	69,424	97.25	93.30	3.95
2004	71,223	94.43	93.29	1.14
2005	73,070	91.61	93.27	−1.66
2006	74,966	92.22	93.26	−1.03
2007	76,910	92.84	93.24	−0.41
2008	78,906	93.45	93.23	0.22
2009	80,954	94.07	93.21	0.85
2010	83,055	90.85	93.20	−2.35
2011	85,211	87.63	93.18	−5.56

Table 19.7 presents the forecasted figures of the forest coverage (as percentage of total land area) on the basis of the estimated relationship presented in Table 19.6. The absolute value of forecasting error ranges from 0.22 in 2008 to 5.56 in 2011. Considering the magnitudes of errors, which are all less than 6 % throughout the period, it can be concluded that the log-linear model satisfactorily fit the changes in forest cover on account of population pressure.

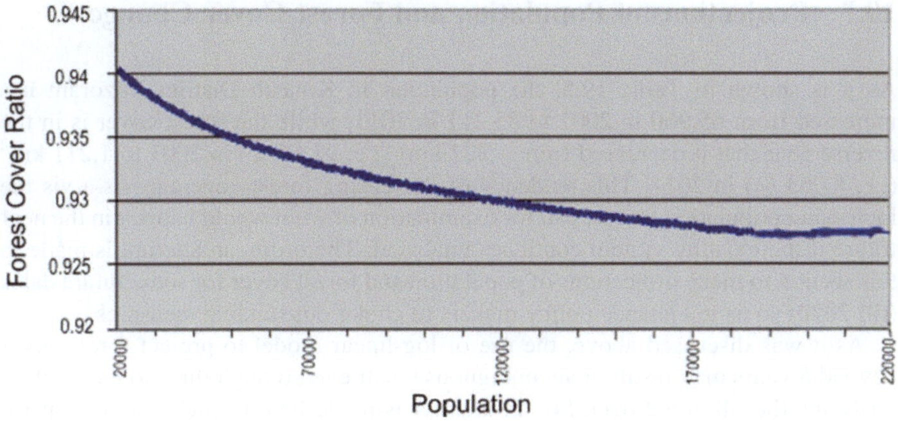

Fig. 19.3 Forecasted values of forest cover of the estimated log-linear model

However, a closer look at the trend of the estimated error whose magnitude systematically decreases with the increase in population till the population figure reach 78,906 in 2008; and increases systematically afterward. This is due to the convexity of the function (if $\beta < -0.1$) resulting in the systematic increase of forecasting error beyond 2008. Therefore, if the explanatory variables or predictor increases at a compound growth rate, the log-linear model is not, theoretically, suggested for forecasting forest cover for a longer period (i.e. more than 5 years) as the absolute value of the percentage of error are likely to be more than 10 % five years hence; even though this model satisfactorily fits the existing time series data of population and forest cover. The forecasted percentage of forest cover ratio on the basis of estimated log-linear model given in Table 19.6 is presented in Fig. 19.3.

The case of slower rate of deforestation after some point of the population growth may be taken in the other way that population growth may not be the significant factors of deforestation in the study area beyond this point. As it is pointed out in the preceding section that economic growth results in the shifting of the population from traditional livelihood of the practice of jhumming and other forest related activities to a better and sustainable livelihood, especially in service sector can, inversely, result into environmental improvement. This case is hypothesized by Kuznets (1955) in what is later known as 'Environmental Kuznet Curve' (EKC). EKC states that in the early stages of economic growth environmental degradation and pollution increase, but beyond some level of income per capita the trend reverses, so that at high-income levels economic growth leads to environmental improvement.

19.7 Projections of Population and Forest Cover Change

As it is shown in Table 19.5, the population in Kolasib District Mizoram has increased from 65,960 in 2001 to 85,211 in 2011; while the forest cover is in the reverse gear that it decreased from 1,322 km^2 (i.e. 95.66 %) in 2001 to 1,211 km^2 (i.e. 87.63 %) in 2011. This tendency of decreasing forest coverage vis-a-vis the increasing population trends, call for examination of what would happen in the near future if the existing system continues unabated. Therefore, an attempt is made in this section to make projections of population and forest cover for some future dates (till 2020) so as to enhance policy makers to choke out suitable measures.

As it was discussed above, the use of log-linear model to project forest cover beyond 5 years may result in an ambiguous result even though this model satisfactorily fits the observed data. So, an attempt is made here to make projection for population in the study area as well as forest cover percentage till 2020 by estimating the rate of growth of the two variables. Therefore, the projections made in this section will be based on the growth rate of each of the relevant variables rather than their functional relationship. The analysis adopt the decadal population figures of population in Kolasib District from 1981 to 2011 (Table 19.8) and the forest cover data obtained from Forest Survey of India (Table 19.5). The compound annual growth rate (CAGR) of population has been worked out from the following the growth curve:

$$\begin{aligned} P_t &= P_o(1+r)^t \\ logP_t &= logP_o + bt \end{aligned} \tag{19.5}$$

where P_t is the population at time t, P_o is the initial population and $b = log\ (1 + r)$; while $r = e^b–1$ is the compound growth rate. The estimated result of Eq. (19.5) is presented in Table 19.9.

Table 19.9 reveals that the growth curve as proposed in Eq. (19.5) fits the observed population data very well that the R-square is 0.989, indicating that we can consistently use the CAGR worked out from this growth curve for projecting population in Kolasib district.

The same growth curve is estimated to work out the compound annual growth rate (decay) of forest cover by changing population for forest cover as in Eq. (19.5). However, the data set used in this case will be the annual forest cover data in Table 19.5, rather than decadal census data. This is due to an unavailability of the relevant information on forest cover corresponding to these decadal dates. The Estimated growth curve of forest cover is presented in Table 19.10. Table 19.10

Table 19.8 Population growth in Kolasib district of Mizoram

Cases	1981	1991	2001	2011
Population	6,488	6,142	9,547	13,179
Number of household	1,042	1,099	2,075	2,717
Average family size	7	6	5	5

Source: Directorate of Census Operation, Mizoram, Aizawl

Table 19.9 Estimated population growth curve

Dependent variable: LOG(P_t)				
Method: Least squares				
Variable	Coefficient	Std. error	t-Statistic	Prob.
Constant	8.508306	0.049187	172.9776	0.0000
Time	0.245880	0.017961	13.68990	0.0053
R-squared	0.989441	Mean dependent var		9.123005
Adjusted R-squared	0.984162	S.D. dependent var		0.319119
S.E. of regression	0.040161	Akaike info criterion		−3.284974
Sum squared resid	0.003226	Schwarz criterion		−3.591827
Log likelihood	8.569948	Hannan-Quinn criter.		−3.958340
F-statistic	187.4134	Durbin-Watson stat		3.317900
Prob(F-statistic)	0.005293			
Compound Annual Growth Rate (CAGR): 2.49 %				

Table 19.10 Estimated growth curve of forest cover

Dependent Variable: LOG(FOREST)				
Method: Least Squares				
Sample: 2001 2011				
Included observations: 11				
Variable	Coefficient	Std. Error	t-Statistic	Prob.
Constant	−0.026764	0.012185	−2.196509	0.0557
Time	−0.007140	0.001797	−3.974256	0.0032
R-squared	0.637019	Mean dependent var		−0.069605
Adjusted R-squared	0.596688	S.D. dependent var		0.029670
S.E. of regression	0.018843	Akaike info criterion		−4.942423
Sum squared resid	0.003195	Schwarz criterion		−4.870078
Log likelihood	29.18333	Hannan-Quinn criter.		−4.988026
F-statistic	15.79471	Durbin-Watson stat		1.211866
Prob(F-statistic)	0.003234			
Compound Annual Growth Rate: −0.698 %				

shows the estimated coefficient is significant at 1 % and the estimated compound growth rate turned out to be −0.698 %. This result indicates forest cover in Kolasib district is expected to decrease by almost one per cent (0.698 %) every year and this must be a serious concern of the various stakeholders.

Assuming the annual growth rate (worked out from decadal growth rate) of population from 1981 to 2011 and growth rate of forest cover from 2001 to 2011 to hold till 2020, the study made projections of population and forest area in Kolasib District till 2020, and are presented in Table 19.11. This table projected that forest area cover of the district decreases by 327.4 km^2 during the period of 10 years only; while population is projected to increase by as much as 1,72,518 persons.

Table 19.11 Population projections and projected forest cover in Kolasib district, Mizoram

Year	Population	Forest cover (km^2)	Forest cover (%)
2011	85,211	1,211.00	87.63
2012	87,333	1,202.55	87.01
2013	91,736	1,185.82	85.80
2014	98,761	1,161.16	84.02
2015	108,971	1,129.08	81.70
2016	123,230	1,090.22	78.89
2017	142,826	1,045.35	75.64
2018	169,659	995.33	72.02
2019	206,552	941.09	68.10
2020	257,729	883.60	63.94

19.8 Concluding Remarks

The North-East Region of India is one of the recognized global biodiversity hotspots; approximately 30 % of total forest cover is under pressure of rapid land use changes. This region harbors variety of rare and endemic species of flora and fauna. It also has a strong bearing on regional climatic conditions. Extensive shifting cultivation, compounded by increasing population pressure and demands for agriculture land are the prime drivers in addition to other proximate drivers of deforestation. Kolasib district is the only district witnessing deforestation among the eight district of Mizoram at the rate of 7.31 % between 2001 and 2011, while the rest of seven districts show significant improvement in forest cover at the same assessment year. This signifies the relevance of selecting Kolasib district as a study area.

The sample survey data analysis shows that shifting agriculture has significantly decreased both in very accessible and medium accessible villages. However, it is increasing in inaccessible villages. The total forest cover has increased slightly in accessible villages, but declined significantly in inaccessible villages. Forests seem to be under pressure, both from population increase as well as rising aspirations for economic development. Dense forests and traditionally managed safety/supply reserves have both decreased in all strata in a significant manner with maximum change in inaccessible villages. Partly, dense forest might have degraded into open forests on account of overuse; whereas in inaccessible villages forest has given place to shifting agriculture. The most unwelcome change is loss of safety and supply reserves forest across all the villages.

The land management and ownership issue in Kolasib district is complex. Land is temporarily allotted to individual for the period of 5 years. But, the individual often illegally claim the land permanently even after 5 years. The state Government Policy like New Land Use Policy (NLUP) encouraged land entitlement to individual depending on the trade scheme. Once the land is allotted to individual under the said policy, the ownership is often claimed by individual even if the land is or is not utilized for the purpose trade scheme. This had encouraged conversion of forested land to agriculture, horticulture and plantation in the claimed private land.

The main causes of deforestation in Kolasib district can be classified as proximate and underlying causes. The population, household and livelihood changes are the main socio-economic variables identified during sample survey. They are also acting as the proximate causes along with shifting cultivation and plantation. Various state government policies like New Land Use Policy, absence of proper land management and accessibility coupled with market access played a pivotal role as underlying causes of deforestation in the district.

The analysis by the model adopted observed that the estimated coefficient is significantly negative indicating that changes in the population and household is negatively responded to by forest cover while livelihood changes is positively responded to by improvement in forest cover. So, it may be concluded that the Logarithmic Model being proposed here to fit the forest land use change is significantly valid in case of time series as well as cross section data; and hence it may be considered as appropriate model to fit the observed forest land use change.

References

Ackoff RL, Gupta SK, Minas JS (1962) Scientific methods optimising research decisions. Wiley, New York

Ball JB (2001) Global forest resources: history and dynamics. In: Evans J (ed) The forests handbook, vol 1. Blackwell, Oxford, pp 3–22, 418 pp

Barraclough SL, Ghimire KB (2000) Agricultural expansion and tropical deforestation: poverty, international trade and land use. Earthscan, Sterling, Virginia

Census of India (1981) Directorate of census operation, Mizoram, Aizawl

Census of India (1991) Directorate of census operation, Mizoram, Aizawl

Census of India (2001) Directorate of census operation, Mizoram, Aizawl

Cochran WG (1963) Sampling techniques, 2nd edn. Wiley, New York, pp 87–94

FAO (1993) Forests resources assesment 1990: Tropical countries. FAO Rome, Italy

FAO (2001) FAO statistical databases. United Nations (http://apps.fao.org)

FSI (1999) The state of forest report. Forest Survey of India, Dehradun

FSI (2001) State of forest report. Forest Survey of India, Ministry of Environment and Forests, Dehradun

FSI (2003) State of forest report. Forest Survey of India, Ministry of Environment and Forests, Dehradun

FSI (2005) State of forest report. Forest Survey of India, Ministry of Environment and Forests, Dehradun

FSI (2009) State of forest report. Forest Survey of India, Ministry of Environment and Forests, Dehradun

FSI (2011) State of forest report. Forest Survey of India, Ministry of Environment and Forests, Dehradun

Hussain M (2007) Models in Geography. Rawat publication, Jaipur, p 36

Kuznets S (1955) Economic growth and income inequality. Am Econ Rev 49:1–28

Lambin EF (1997) Modelling and monitoring land-cover change processes in tropical regions. Progr Phys Geogr 21:375–393

Lambin F, Geist HJ, Lepers E (2003) Dynamics of land use land cover change in tropical regions. Annu Rev Environ Resourc 28:205–241

Mittermeier R, Mittermeier CG, Gil PR, Pilgrim J, Fonceca G (2003) Wilderness: Earth's last wild places. Univ. Chicago Press, Chicago, p 576

Myers N (1994) Tropical deforestation: rates and patterns. In: Brown and Pearce (eds) The causes of tropical deforestation. University College London Press, United Kingdom, pp 27–40
Singh KD, Jans K (1995) Assessing the world' forest resources. Nat Resour 31(2):32–40
Singh RL, Singh PB (1991) Elements of practical geography. Kalyani Publishers, Ludhiana, p 238
Singh KD, Sinha B, Lalchamreia H (2012) Technique of survey and planning for conservation and sustainable use of biodiversity in Mizoram, MoEF Funded Research Project Report

Chapter 20
Degradation of Land and Forest Resources: The Story of Shifting Cultivation and Loss of Biodiversity in North-East India

R. Zonunsanga, Ch. Udaya Bhaskara Rao, and P. Rinawma

Abstract One of the relevancies of the Northeast India for scientific investigations has been its richness in biodiversity but the foremost important germane arises from degradation of the same, especially soil resources, due to the existence of fragile ecosystems which include combinations of hostile physical and cultural environments within the hilly forested milieu of the tribal people. Great efforts have been made by agencies for implementation of alternative farming systems to do away with the traditional shifting cultivation to menace rapid rate of degradation to resources, but have failed. An attempt has been made to quantify soil loss from the hilly regions using a mathematical model and efforts have been rendered to measure the spatial extent under degradation of resources and different land use systems with the applications of advanced techniques. Special attention has been given to shifting cultivation and measures to control soil erosion have been suggested which could be socially acceptable and economically affordable by the poor tribal farmers in hilly forested terrain.

Keywords Degradation • Resources • Shifting cultivation • Soil erosion • Tribal

20.1 Introduction

Anthropogenic activities of the tribal people in the hilly regions have largely resulted to environmental degradation due to the prevailing unfavourable physical conditions. In most cases, academicians, bureaucrats and planners including policy makers have attributed the system of shifting cultivation as the root-cause of degradation to land and forest resources. There exists no doubt that this agricultural practice has indeed

R. Zonunsanga (✉) • Ch.U.B. Rao • P. Rinawma
Department of Geography and Resource Management, School of Earth Sciences, Mizoram University, Aizawl, Mizoram, India
e-mail: cocas.2008@rediffmail.com; chegondiu@yahoo.co.in; ascassdir@gmail.com

M. Singh et al. (eds.), *Climate Change and Biodiversity: Proceedings of IGU Rohtak Conference, Vol. 1*, Advances in Geographical and Environmental Sciences,
DOI 10.1007/978-4-431-54838-6_20, © Springer Japan 2014

caused countless negative consequences out of which deforestation gained important priority for being the most noticeable or visible effect. However, if one considers the necessities of cultivation for food and other basic needs, as well as the economic and social conditions of the forest dwellers, it would be wrong to suggest and recommend the abolition of shifting cultivation for being the aged-old tradition and more importantly the only primary source of livelihood. Being under such compulsion to practice agricultural system in the rugged hilly terrain, where scarcity of land resources pose a serious threat to food security, farmers have to keep on with this traditional land culture system practiced along the steep hill slopes. The great efforts rendered by the Government in trying to do away or implement alternatives to the shifting cultivation, in most cases, therefore proved failures.

Further, if one takes into account the multifacets of deforestation along with other allied environmental degradations and their causes, the prevalent shifting cultivation system, as the causes of forest clearance, comes as secondary factor whereas the primary root-cause hails from soil erosion so far as the farmers of N.E India are concerned. This is true because, soil erosion induces reduction of soil fertility and moisture retention capacity of soils that farmers have to shift their agricultural lands from one place to another hence deforestation and other allied resources degradation. The best way to dwindle the system of shifting cultivation therefore comes with measures of controlling the loss of soil to regulate the compulsion of shifting the agricultural lands.

20.2 Background of the Study Area

The present research works have been confined to the watershed of River Teirei sandwiched between the hill ridges of Hachhek and Mamit, covering an area of 68,000 ha to represent the northeast India where shifting cultivation has been extensively practiced by the forest dwellers. The area is located at the north-western part of the Mizoram state and lies, geographically, between 23°54′66″N and 24°23′56″N latitudes and between 92°34′53″E and 92°54′91″E longitudes. The Survey of India topographical sheets viz. 83D/8, 83D/12, 84A/5, 84A/6 and 84A/9 cover the study area stretching in a north-south direction.

Geologically, the area belongs to the Surma group with the hills consisting of sandstones and shales of Tertiary age, thrown into long folds (Pachuau 1994). It is the southern extension of Purvanchal of the Himalayas, formed by the accumulated sediments of the then existing Tethys Sea, uplifted by compressional tectonic forces of the Indian plate. It is composed predominantly of mountainous terrain, the ranges which are inclined in north to south direction in parallel series, separated by narrow deep river valleys. The maximum elevation of 1,095 m (Above MSL) is found at the southern-most part which serves the upper course and source of the river Teirei and the lowest with 22 m (Above MSL) at the northern-most part of the watershed where the river Teirei confluences with the river Tut.

Meteorologically, the area, owing to its tropical-monsoonal location experiences precipitous characteristics with high humidity and sufficient solar radiation which are modified by elevations of local hilly terrains. The average annual rainfall during the year 1987–2011 is 2,882 mm and the average minimum and maximum temperature 15.02 °C and 31.65 °C respectively with an average annual humidity of 70 %. Rainfall is caused by the south-west monsoon and thus confined its occurrence generally during April to September.

The watershed accommodates 34 numbers of different types of settlements such as hamlets, villages and towns with a total population of approximately 33,000 persons. There are about 6,900 households or families out of which 73 % is estimated to belong to Jhummias or shifting cultivation for livelihood. An area of 5 km^2 (approximately) within the study area is used every year for shifting cultivation allotted generally from the villagers' owned community lands. This community lands account for almost 40 % of the watershed total area which means that the total potential areas for shifting cultivation is as large as 272 km^2 provided the location of the community lands is within the viable distance for access from the villages. This type of cultivation, the traditional system still practiced popularly in the area is, economically and ecologically, the most unsustainable system of cultivation leading to destruction of valuable natural forests through the processes of slashing, burning and abandoning the cultivated lands.

The extensive utilization and practice of this primitive agricultural system—the jhumming and the rapid uncontrolled increase in population have been exerting tremendous pressure and degradation on the available natural resources. The area has a population density of 48.53 persons/km^2 which is high enough for areas with mountainous terrain where favourable culturable land is limited. The carrying capacity of the existing permanent agricultural land within the study area is estimated as 1,320 persons/km^2 as against the figure of 1,130 in the district as a whole.

20.3 Materials and Methods

The root-cause of degradations to land, water and forest resources being soil erosion in hilly terrain of North-east India, estimations of soil loss from each spatial unit in the study area have been performed through integrations and overlaying of different GIS generated thematic layers such as rainfall erosivity, soil erodibility, slope length and steepness, land use and vegetation/canopy cover and, anthropogenic management practices. Special attention has been given to shifting cultivation in regard to its areal extent and contribution towards soil erosion. The Revised Universal Soil Loss Equation with minor modifications has been applied to quantify the metrological loss of soil within the study area that computations of the value of each factor controlling the erosion have been performed to derive at the quantifiable soil loss on a given unit area.

20.4 Soil Erosion and Land Degradation

Soil erosion is a process, complex and dynamic, by which soil surfaces are detached, transported and accumulated in a distant place. It is a serious threat to the long term sustainability of land resources throughout the Third World. Its impact affects not only land productivity but also the environment downstream by sedimentation of the eroded particles, thereby decreasing the carrying capacity of water bodies. Scientifically, the main attributor to land degradation is soil erosion by runoff water (Angima 2003). One of the most unsustainable agricultural system—shifting cultivation has also been attributed as the outcome of severe soil erosion whereby the farmers have been compelled to shift due to intensive loss of the fertile topsoil. Of the World's land degradation problem, soil erosion is the first order category (Hitzhusen 1993). Conservation and management practices of land resources should therefore consider minimizing soil erosion.

The largest land degradation category in India is the land affected by water erosion (Venkataratnam and Ravisankar 2004). The entire study area is also afflicted with a serious problem of soil loss as a consequence of its negative characteristics in regard to erosivity, erodibility, topography, vegetation covers and absence of management practices. Based on the estimated annual soil loss per hectare, the watershed of River Teirei is classified into 6 (six) erosional intensity units. The computed soil loss data and final erosion intensity map, in general and interestingly, follows the pattern of slopes distribution and more significantly the pattern of vegetation density. This shows that the rate of soil erosion in hilly terrain is highly associated and determined by the protective services of plants and trees. Out of the total area of the watershed, about 14,225 ha (21 %) falls under very slight intensity zone of soil erosion (0–10 t/ha/year) and the zone with slight intensity of soil erosion (10–20 t/ha/year) covers about 13,300 ha (19.5 %) of the watershed. These two zones covering about 40 % of the watershed fall under the category where erosion-induced land degradation risk is low. The soil cover being deep to very deep and the vegetative cover being thick in the area, chances exist for soil resilience and capability to regain their degraded energy.

The area with the largest areal extent is the moderate intensity zone of soil erosion (20–40 t/ha/year) accounting for 37.65 % of the river basin. The severe intensity zone of erosion (40–80 t/ha/year) shares approximately 4.9 % and the erosion intensity zone with soil loss above 80 t/ha/year, designated as 'very severe' covers an area of about 14.9 % (10,125 ha) of the watershed. These three zones accounting for about 60 % of the total area fall under the threatening category with high risk to land degradation and loss of biodiversity. The average annual soil loss in this area is as high as 112 t/ha revealing the possibility of complete top soil removal in about 20 years if no conservation measures are implemented. The area had indeed ranked the topmost priority for conservation of soil in the state of Mizoram (IRDAS 1994)

20.5 Shifting Cultivation and Degradation of Resources

Shifting cultivation is largely a subsistence activity practiced in the area where few alternative options exist due to physico-socio-economic constraints and its practice is therefore likely to continue. The paraphrasing of the term 'shifting cultivation as necessary evil' has great demonstrative value for policy framers (Das 2006) and even academicians because the system is attributed as the root-cause of degradation to land and forest resources. Looking at the odd side of a coin, there exists countless negative consequences of this agricultural practice. This agricultural system covers an area of about 5 % of the total area annually and the abandoned lands devoid of proper vegetation cover account for approximately 15 % every year. These abandoned lands after few years generally regenerated to less dense forest dominated by bamboo whose areal extent figures about 56 % of the river basin. This revealed that absolute deforestation or complete degradation of forest resources due to shifting cultivation has been experienced in 5 % of the total area annually and that the absolute but temporal degradation of forests resources has been witnessed on 20 %. These rates at which spatial reduction in forest cover occur are quite detrimental to the fragile Himalayan ecosystem and biodiversity (Table 20.1).

Taking into account the very severe intensity zone where annual loss of soil cover per hectare exceeds 80 t, soil erosion is in operation actively regardless of the soil types, slopes and rainfall but vegetation cover only. As high as 96.3 % of the area under this category of very severe erosion intensity come under shifting cultivation thereby manipulating the chances of biodiversity loss. In most cases, jhumming is largely associated with moderately sloping—steep slopes and under a situation where jhumming operate together with very steep slopes, soil loss exceeds even 900 t/ha/yr. Under such conditions, removal of all the soil nutrients, moisture retention capacity and fertility is obviously certain as no conservation measure is

Table 20.1 Soil erosion intensity and land degradation (*Source*: Author/Self)

Erosion intensity classes	Soil loss classes (ton/ha/year)	Erosion-induced land degradation risk	Area (ha)	Area (in %)
Very slight	00–10	Very low risk to land degradation	14,225	20.90
Slight	10–20	Low risk to land degradation	13,300	19.56
Moderate	20–40	High risk to land degradation	25,600	37.65
Severe	40–80	High risk to severe land degradation	3,325	04.89
Very severe	80 above	Very high risk to severe land degradation	10,125	14.89
Unclassified	Dynamic	Built-up lands	1,425	02.10

implemented that the subsistent farmers have been compelled to shift their agricultural lands so as to maintain their annual productivity. The shift from one plot to another later became traditional practice.

Thus, the prevalent shifting cultivation system, as attributed to the causes of degradation to biodiversity, land and forest resources, remains only as secondary factor whereas the primary root-cause hails from soil erosion. The best way to manage and reduce the system of shifting cultivation is therefore regulation of soil loss from the farmland to avoid the compulsion of shifting the agricultural lands because several attempts have been made in the northeast India to do away with this agricultural system but have failed.

Since no measure (mechanical, technical etc.) appears to be affordable or viable to control or abolish the so-called shifting cultivation due to the prevailing physico-socio-economic conditions, only modification and improvements measures to agricultural practice could reduce the soil erosion rate. Such measures for improvement, economically and culturally viable, include the leasing of trees with thick and large canopy while slashing or burning and also linear patches of the most abundant fauna (usually bamboos) across the slope to act as natural contours to reduce runoff velocity and to increase residue (Figs. 20.1 and 20.2).

20.6 Conclusion

The fragile ecosystem of hilly terrain in Northeast India has been experiencing a threatening and high risk of soil erosion-induced land degradation estimated to cover about 60 % of area under investigation. The very high rate of deforestation (5 % and 15 %, fresh and 3 years fallow of the total area respectively every year) has worsen the situation exerting tremendous pressure to degradation of forest resources further resulting to loss of biodiversity. Shifting cultivation is therefore attributed as the major contributor towards the three negative scenarios, i.e. soil erosion-induced land degradation, forest degradation and loss of biodiversity. It is, however, to be documented that the system of shifting cultivation cannot be simply paraphrased as "necessary evil" for it is the age-old and blood-bonded traditional practice that occupies the heart of tribal culture which is, arguably, one of the most environmental friendly practices. Besides, uncountable constraints exist towards any attempt to do away with the system that the only viable action to control degradation of biodiversity, land and forest resources comes from the implementation of improvement measures to shifting cultivation system so that farmers retain the soil nutrients in their farmland and avoid shifting from one plot to another.

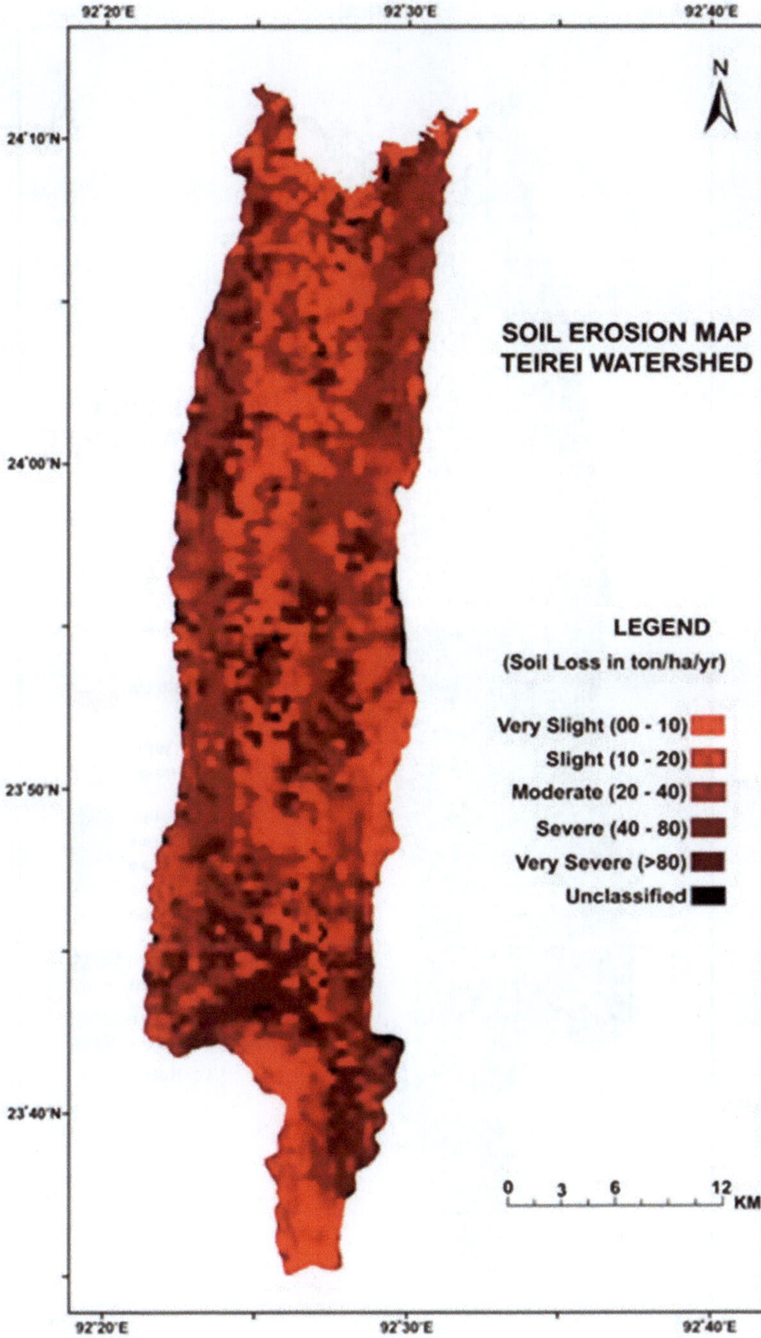

Fig. 20.1 Spatial distribution of soil erosion (*Source*: Author/Self)

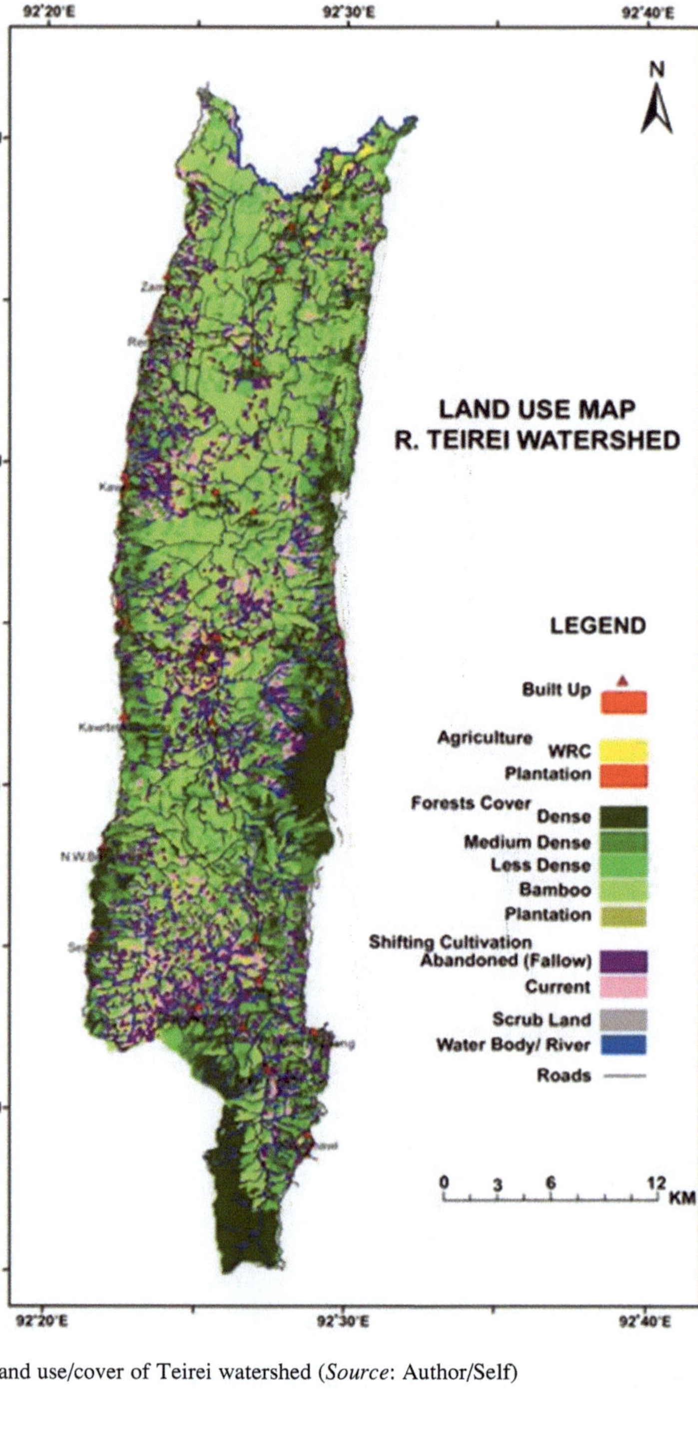

Fig. 20.2 Land use/cover of Teirei watershed (*Source*: Author/Self)

References

Angima (2003) In A.H. Sheikh's "Integration of GIS and USLE for soil loss estimation in a Himalayan watershed". Recent Res Sci Technol 3:51–57

Das D (2006) Demystifying the Myth of shifting cultivation agronomy in the North-East, Economic and Political Weekly, November 25, 2006

Hitzhusen (1993) In A.H. Sheikh's "Integration of GIS and USLE for soil loss estimation in a Himalayan watershed". Recent Res Sci Technol 3(3):51–57

IRDAS (1994) Soil erosion and land degradation problems in Mizoram, Institute of Resource Development and Social Management, Hyderabad, Govt. of Mizoram sponsored Project

Pachuau R (1994) Geography of Mizoram. RT Enterprise, Aizawl

Venkataratnam L, Ravisankar T (2004) Remote sensing for soils and land degradation. In: Gautam NC, Raghavswamy V (eds) Land use/land cover and management practices in India. B.S. Publication, Hyderabad

MIX
Papier aus verantwortungsvollen Quellen
Paper from responsible sources
FSC® C105338

If you have any concerns about our products,
you can contact us on
ProductSafety@springernature.com

In case Publisher is established outside the EU,
the EU authorized representative is:
Springer Nature Customer Service Center GmbH
Europaplatz 3, 69115 Heidelberg, Germany

Printed by Libri Plureos GmbH
in Hamburg, Germany